Moisture Sorption in Wood Cell Walls
木材细胞壁水分吸附

Wu Yiqiang　Guo Xin

吴义强　郭　鑫　著

Science Press

Beijing

Responsible Editor: Huo Zhiguo

Copyright © 2020 by Science Press
Published by Science Press
16 Donghuangchenggen North Street
Beijing 100717, China

Printed in Beijing

All rights reserved. No part of this publication may be reproduced, stored in a retrieval system, or transmitted in any form or by any means, electronic, mechanical, photocopying, recording or otherwise, without the prior written permission of the copyright owner

ISBN 978-7-03-066058-9 (Beijing)

Preface

As the renewable and recycled timber materials among the world's four major materials (steel, cement, wood, plastic), wood plays an irreplaceable and even more and more important role in the people's livelihood. However, wood is a hygroscopic material, and its dimension and density, as well as its mechanical, elastic, and electrical properties are affected by its moisture content. A fundamental understanding of wood-water relationship is very important with respect to the wood's commercial utilization as well as its biology and chemistry. This is the reason why the wood-water relationship has been studied since the beginning of wood research. There are over thousand representative publications available which have provided some useful information about the wood-water relationship. However, due to the complexity of chemical constituents and the heterogeneity of anatomical structure of wood, there are still many issues worthy of further study such as the distribution and existing states of water in wood cell walls. With the development of new technology, vast modern analytical methods are emerging, and the study of the wood-water relationship is changing from macro to micro scale and from cell to molecular level.

In the present book, the authors attempt to give summaries on their recent works about moisture sorption in wood cell walls. In Chapter 1, the authors review the recent analytical techniques which have been applied to demonstrate the distribution and existing states of water in wood cell walls, and show the advantage of micro-FTIR and confocal Raman spectroscopy in these research directions. From chapter 2 to chapter 5, a new approach for quantitative evaluation of moisture sorption in nanogram quantities of wood and its major chemical components based on the micro-FTIR spectroscopy is developed. From chapter 6 to chapter 9, molecular associations of water with wood and its major chemical components are studied and a possible moisture adsorption mechanism is raised. In chapter 10, the spatial distributions of water and the major chemical components of wood are determined by micro-FTIR spectroscopy. In chapter 11 and chapter 12, the water vapor sorption properties of the major chemical components of wood are studied by dynamic vapor sorption apparatus.

The authors hope that these studies not only establish the theoretical system of the moisture sorption of wood at the cellular and molecular level, but also provide theoretical guidance to improve these technologies of wood modification, wood drying and preparing the high-value wooden material.

The authors express their sincere thanks to the supports from Major Program of National Natural Science Foundation of China (31890771), Key Program of National Natural Science Foundation of China (31530009), National & Local Joint Engineering Research Center of Green Processing Technology for Agriculture and Forest Biomass, Provincial and Ministerial Collaborative Innovation Center for High-Efficiency Utilization of Wood and Bamboo Resources and Hunan Key Laboratory of Biomass Materials and Green Conversion Technology. The authors also extend their gratitude to Prof. Wu Qinglin , Prof. Yan Ning, Prof. Xie Xinfeng, Prof. Qing Yan, and other contributors for their precious time paid to this book.

Contents

Chapter 1 Overview 1
  1.1 Introduction 1
  1.2 Characterizing the spatial distribution of adsorbed water in wood 3
    1.2.1 Magnetic resonance imaging technique 3
    1.2.2 Computed tomography scanning technique 4
    1.2.3 Neutron radiography 5
    1.2.4 Vibrational spectroscopic imaging techniques 6
  1.3 Determining molecular interactions between adsorbed water and wood 7
    1.3.1 Near-infrared spectroscopy 7
    1.3.2 Nuclear magnetic resonance technique 8
    1.3.3 Fourier transform infrared spectroscopy 9
    1.3.4 Raman spectroscopy 10
  1.4 Future directions 11
  References 11

Chapter 2 Quantitative detection of moisture content in heat-treated wood cell walls using micro-FTIR spectroscopy 17
  2.1 Introduction 17
  2.2 Materials and methods 20
    2.2.1 Materials 20
    2.2.2 Micro-FTIR spectrometer 20
    2.2.3 DVS apparatus 21
    2.2.4 Micro-FTIR data processing 22
  2.3 Results and discussion 23
    2.3.1 Qualitatively analyzing moisture sorption 23
    2.3.2 Quantitative analysis of moisture sorption 26
  2.4 Conclusions 28
  References 29

Chapter 3 Quantitative analysis of moisture sorption in lignin using micro-FTIR spectroscopy 34
  3.1 Introduction 34

3.2 Materials and methods ·· 37
  3.2.1 Materials ··· 37
  3.2.2 Experimental apparatus for micro-FTIR spectral measurement ······· 37
  3.2.3 Experimental apparatus for moisture content measurement ············ 38
  3.2.4 Micro-FTIR spectral data processing ·································· 39
3.3 Results and discussion ··· 40
  3.3.1 Quantitative analysis of moisture adsorption in lignin ····················· 40
  3.3.2 Quantitative evaluation of moisture adsorption in lignin ················ 43
3.4 Conclusions ··· 46
References ·· 46

Chapter 4 Quantitatively characterizing moisture sorption of cellulose using micro-FTIR spectroscopy ·· 51
4.1 Introduction ·· 51
4.2 Experiment section ·· 53
  4.2.1 Sample preparation ·· 53
  4.2.2 Micro-FTIR spectroscopy apparatus ·································· 54
  4.2.3 DVS apparatus ·· 56
  4.2.4 Spectral data processing ·· 57
4.3 Results and discussion ·· 57
  4.3.1 Qualitatively analyzing water adsorption of cellulose nanofiber film ·· 57
  4.3.2 Quantitative analysis of water adsorption in cellulose nanofiber film ·· 60
4.4 Conclusions ·· 63
References ·· 63

Chapter 5 Quantitative evaluation of moisture sorption in TEMPO oxidized cellulose using micro-FTIR spectroscopy ······················· 69
5.1 Introduction ·· 69
5.2 Materials and methods ·· 72
  5.2.1 Materials ·· 72
  5.2.2 Experimental apparatus for micro-FTIR spectroscopy measurement ·· 72
  5.2.3 Determination of moisture content using DVS apparatus ············· 74
  5.2.4 Data processing of micro-FTIR spectra ······························· 75
5.3 Results and discussion ·· 75
  5.3.1 Quantitative analysis of moisture adsorption in TOCNF ············· 75

         5.3.2 Quantitatively evaluating water adsorption of TOCNF ········ 78
    5.4 Conclusions ·· 80
    References ·· 81
Chapter 6 Molecular association of water with wood cell walls during moisture desorption process examined by micro- FTIR spectroscopy ·· 87
    6.1 Introduction ·· 87
    6.2 Materials and methods ··· 90
         6.2.1 Materials ·· 90
         6.2.2 Experimental instrument for micro-FTIR spectral measurement ······ 90
         6.2.3 Micro-FTIR spectral data processing ····················· 91
    6.3 Results and discussion ·· 92
         6.3.1 Effective water sorption sites of wood ···················· 92
         6.3.2 Molecular structure change of water during moisture desorption process ·· 94
    6.4 Conclusions ··· 98
    References ·· 98
Chapter 7 Molecular association of water with heat-treated wood cell walls during moisture adsorption process examined by micro-FTIR spectroscopy ··· 104
    7.1 Introduction ··· 104
    7.2 Experimental section ·· 107
         7.2.1 Sample preparation ··· 107
         7.2.2 Micro-FTIR spectroscopy equipment ··················· 108
         7.2.3 Data processing ··· 109
    7.3 Results and discussion ·· 110
         7.3.1 FTIR spectra of the heat-treated wood associated with water molecules ·· 110
         7.3.2 The analysis of difference spectra ······················· 112
         7.3.3 Intermolecular interactions between adsorbed water and the heat-treated wood ·· 113
    7.4 Conclusions ··· 121
    References ·· 121
Chapter 8 Molecular association of adsorbed water with heat-treated wood cell walls during moisture desorption process examined by micro-FTIR spectroscopy ··· 127

8.1 Introduction ··· 127
8.2 Materials and methods ·· 129
　8.2.1 Materials ··· 129
　8.2.2 Experimental instrument for spectral measurement ···················· 130
　8.2.3 Micro-FTIR spectral data processing ·· 131
8.3 Results and discussion ·· 131
　8.3.1 Effective water sorption sites of heat-treated wood ···················· 131
　8.3.2 Molecular structure change of water ·· 133
8.4 Conclusions ·· 137
References ·· 137

Chapter 9 Molecular association of adsorbed water with cellulose during moisture adsorption process examined by micro-FTIR spectroscopy ·· 143
9.1 Introduction ·· 143
9.2 Materials and methods ·· 146
　9.2.1 Materials ··· 146
　9.2.2 Micro-FTIR spectroscopy setup ··· 147
　9.2.3 Data processing ··· 149
9.3 Results and discussion ·· 149
　9.3.1 Micro-FTIR spectra of cellulose nanofiber film ························· 149
　9.3.2 Difference spectra of cellulose nanofiber film at various RH levels ··· 150
　9.3.3 Different types of water adsorbed by cellulose nanofiber film ······ 151
9.4 Conclusions ·· 155
References ·· 155

Chapter 10 Spatial distribution of adsorbed water in cellulose film studied using micro-FTIR spectroscopy ·· 161
10.1 Introduction ·· 161
10.2 Materials and methods ·· 163
　10.2.1 Materials ··· 163
　10.2.2 Micro-FTIR experimental setup ··· 163
10.3 Results and discussion ·· 166
　10.3.1 Qualitatively analyzing water adsorption in cellulose nanofiber film ·· 166
　10.3.2 Spatial distribution of cellulose in the cellulose nanofiber film ···· 167
　10.3.3 Spatial distribution of adsorbed water in the cellulose nanofiber

    film ·· 168
 10.4 Conclusions ··· 171
References ··· 171
Chapter 11 Water vapor sorption properties of sulfuric acid treated and TEMPO oxidized cellulose nanofiber films ······················· 177
 11.1 Introduction ·· 177
 11.2 Material and methods ·· 181
  11.2.1 Materials ··· 181
  11.2.2 DVS apparatus ··· 182
  11.2.3 X-ray diffraction (XRD) ·· 182
  11.2.4 Modulus measurement ·· 183
 11.3 Results and discussion ··· 183
  11.3.1 Water vapor sorption behavior ··· 183
  11.3.2 Sorption hysteresis ··· 185
  11.3.3 Sorption kinetics ·· 187
  11.3.4 The applicability of the Kelvin-Voigt model ································ 189
 11.4 Conclusions ··· 191
References ··· 191
Chapter 12 Water vapor sorption properties of cellulose nanocrystals and nanofibers using dynamic vapor sorption apparatus ················· 197
 12.1 Introduction ·· 197
 12.2 Material and methods ·· 200
  12.2.1 Materials ··· 200
  12.2.2 DVS setup ··· 200
 12.3 Results and discussion ··· 201
  12.3.1 Water vapor sorption behavior ··· 201
  12.3.2 Sorption hysteresis ··· 204
  12.3.3 Sorption kinetics ·· 207
  12.3.4 The applicability of Kelvin-Voigt model ····································· 212
 12.4 Conclusions ··· 218
References ··· 218

Chapter 1 Overview

1.1 Introduction

The development of human civilization has made us obtain unprecedented material and spiritual wealth, which is followed by the increasing demand of human beings for natural resources. As the main body of the ecological environment system, the forest ecological environment system is closely interdependent with the socioeconomic system. One important aspect of this relationship is that forests provide us with wood products. Timber is the only renewable resource among the four basic raw materials (iron and steel, plastic, cement and timber) for the development of the national economy. It is an indispensable "green" raw material resource in the national economic construction and an important means of livelihood to meet the continuous improvement of people's living standards. At present, wood products are widely used in construction, decoration, papermaking, mining, military industry, packaging, transportation, agriculture, energy and other fields, and play an important role in national economic and social development. Wood is one of the important pillars of modern civilization. In the 21st century, with the rapid development of science and technology new materials, some characteristics of wood keep its position as the only raw material available to many architectural structures and furniture products.

Wood is known to be a typical heterogeneous and hygroscopic material (Guindos 2014; Hozjan and Svensson 2011; Olek et al. 2013). The heterogeneity is not only present in anatomical structure but also in distribution of chemical compositions in the cell wall. Microscopically, the wood cell wall is organized in several layers consisted of the primary wall (P), the secondary wall (S), and the middle lamella (ML) (Khalil et al. 2007; Ma et al. 2013; Mellerowicz and Sundberg 2008). The secondary wall can be further differentiated into an outer layer (S1), a middle layer (S2), and an inner layer (S3) (Brandstrom 2001; Lee 1981). The three S layers are distinguishable by the orientation of their cellulose nanofibrils. The ML serves as a cementing layer between the primary walls of the neighboring cells (Agarwal 2006). From a chemical perspective, the wood cell wall contains cellulose, hemicelluloses, and lignin as the main polymers (Hanninen et al. 2011; Karaaslan et al. 2010; Terashima et al. 2009; Zhang et al. 2014). The cellulose molecule is an unbranched polymer of glucose joined by β-1, 4 linkages (Bayer et al. 2006; Li et al. 2017; Peng et al. 2013; Yong and Wickneswari 2013). Cellulose filaments merge in parallel with each other to form microfilaments, which are filled with hemicellulose and pectin. Cellulose in precursor filaments is held together by hydrogen bonds formed by hydroxyl groups between molecules. Van der Waals force also plays a role between cellulose molecules that are close together. Cellulose microfibers are arranged in a regular way to form a clean area, while the other part is called amorphous area. Lignin, the second most abundant component of the wood cell wall (15%-30% by weight), is a cross-linked racemic macromolecule with molecular masses in excess of 10000 Da (Hamad et al. 2012; Lisperguer et al. 2009; Ruiz-Duenas and Martinez 2009). Lignin consists of various types of substructures repeating in a random manner through condensation reactions among three major structural units: coumaryl, coniferyl, and sinapyl alcohols (Hatfield and Vermerris 2001; Ralph et al. 2004; Sakagami et al. 2005). Lignin waterproofs the cell wall enabling transport of water and solutes through the vascular system. Hemicellulose, the third most abundant component of the wood cell wall (15%-25% by weight), is a branched polymer with shorter polymer chains of 500-3000 sugar units as opposed to 7000-15000 glucose units per polymer seen in cellulose (Saha 2003). Hemicelluloses are water- or base-soluble heteropolymers in contrast to the insoluble linear cellulose homopolymer. Meanwhile, the hygroscopicity is referring to that the wood can easily adsorb a large amount of water from hygrothermal conditions of air, and the adsorbed water has a strong influence on the properties of the wood, particularly on the dimensional and mechanical properties (Chauhan and Aggarwal 2004; Evans and Banks 1988; Ishimaru et al. 2001; Molinski and Raczkowski 1988; Tsuchikawa and Siesler 2003). wood has a strong hygroscopicity because the wood contains a large number of hydrophilic groups, and can form hydrogen bonds with water molecules of the group. Hydroxyl and carbonyl groups are the main hydrophilic groups in wood. Because hydroxyl group is stronger than carbonyl group, hydroxyl group is the main research object of wood hygroscopicity. The main components of wood, cellulose, hemicellulose and lignin, have strong hygroscopicity. Here are more hydroxyl groups in the hemicellulose chain of wood, water is easy to enter and moisture absorption is the strongest, while lignin contains a little less hydroxyl group, followed by moisture absorption, most of hydroxyl groups in cellulose crystallization area have formed hydrogen bond and moisture absorption is the least. This shows that heterogeneity and hygroscopicity are closely related. So, the wood-water relationship is very important from the perspectives of both commercial utilization and biology and chemistry of wood. The gravimetric technique is the first introduced method for such investigating adsorption and desorption of water because of its convenience and nicety. There are over thousand representative publications available on this method which has provided vast quantitative data on the amount of adsorbed water in wood. However, due to the extremely heterogeneity of wood, the spatial distribution and molecular association of the adsorbed water in wood are very complicated subjects in the wood-water relationships, which has attracted the attention of many scientists. Therefore, our main objective is to review the recent progress of the study on the spatial distribution and molecular association of the adsorbed water in wood.

1.2 Characterizing the spatial distribution of adsorbed water in wood

Water runs through the whole process of wood formation, with the birth of solid wood out of the tree. Wood has many affinity points that can be combined with water, and there is water exchange between wood and water in different humidity environment. The cell cavities, cell gaps and interstices in the cell walls of wood constitute a complex capillary system, which makes the movement of water in wood in a variety of forms. The spatial distribution of water in wood is very complex, and it has been an important basic research field of wood science. A number of experimental approaches have been advanced to be able to characterize the spatial distribution of adsorbed water in wood.

1.2.1 Magnetic resonance imaging technique

Magnetic resonance imaging uses the principle of nuclear magnetic resonance, because the energy has different attenuation in different material structure, the electromagnetic wave can be detected by external gradient magnetic field. Magnetic resonance imaging (MRI) technique is capable of nondestructive investigation of the wood on both microscopic and macroscopic levels, which can provide detailed and quantitative spatial distribution information on water adsorbed by wood (Brownstein 1980; Casieri et al. 2004; Menon et al. 1987; Merela et al. 2005; Rosenkilde and Glover 2002). For example, Araujo *et al.* obtained one-dimensional images of the bound water and the lumen water in white spruce sapwood separately at a range of moisture contents from 100% to 17% by MRI technique (Araujo et al. 1992). Passarini *et al.* mapped moisture distribution on heterogeneous hardwoods by MRI technique, and interpreted the obtained mapping images with the help of scanning electron microscopy (Passarini et al. 2015). Rosenkilde *et al.* obtained the spatially resolved MRI of the moisture content (MC) in wood with a spatial resolution of 20 mm (Rosenkilde and Glover 2002). Dvinskikh *et al.* demonstrated spatially resolved moisture adsorption into scots pine with a spatial resolution on a sub-millimetre scale by [^{1}H] MRI based on an effective single-point imaging (SPI) sequence (Dvinskikh et al. 2011).

MRI technique can reach a higher accuracy in quantifying the moisture content than X-ray, but still in the range of 2-4 kg/m^3. Wood researchers have not favored the use of MRI, in part due to the relatively expensive equipment. More importantly, it is particularly difficult to measure the moisture content below the fiber saturation point (FSP) in wood with magnetic resonance imaging techniques because of shorter relaxation times (T2). This relaxation time is short for water in wood below the FSP because the water molecules are bound to macromolecules and have the characteristics of a solid. The molecular motional averaging process (which gives liquids their narrow line-widths or long T2 values) is much reduced. T2 characterizes the nuclear signal decay and, taken together with the applied magnetic field gradient, governs the achievable resolution.

1.2.2 Computed tomography scanning technique

Another non-destructive technique to study the water distribution in wood is the computed tomography (CT) scanning based on X-ray or gamma-ray, but with a depth resolution of approximately 240 μm. CT is a completely different imaging method from general radiographic imaging. Generally, radiometric imaging is to project a 3-D object onto a 2-D plane. All the images overlap and interfere with each other. The image is not only blurred, but also loses depth information. But CT can show the structure relationship, material composition and defect condition of the detected part clearly and accurately, and the detection effect is inferior to other traditional NDT methods. Compared with other nondestructive testing techniques, CT has the advantages of strong penetration, high resolution, fast testing speed, intuitive testing results, and no need to contact with the object under test. X-ray radiography and computer tomography of uptake experiments in wood show that water rise is strongly uneven in softwood and depends upon the location of earlywood and latewood rings in the sample. For example, Davis *et al.* measured moisture distribution of mountain ash (Eucalyptus regnans) using the CT scanner based on gamma-ray (Davis et al. 1993).

Fromm *et al.* investigated the water distribution within the sapwood and heartwood of green spruce (Picea abies) and oak (Quercus robur) L. as well as in the earlywood and latewood of an annual ring using X-ray computed tomography with a spatial resolution of 0.1225 mm^3 (Fromm et al. 2001). A disadvantage of X-ray radiography is the limited accuracy of ±13.4 kg/m^3 in quantifying the moisture content.

CT scanning measures give a combination of both wood and water. There is a particular problem of this method that the density of the dry wood sample has to be known in order to calculate the moisture content of subsequent samples.

1.2.3 Neutron radiography

Neutron radiography has been established as a conventional method of NDT, which is similar to X-ray and gamma radiography in the principle of detection. The neutron rays passing through the object can reflect the structure information of the object through the attenuation of intensity to realize the NDT. Neutron radiation has a strong penetration ability to most metal materials, some even far beyond X-ray and gamma-ray, but some materials containing hydrogen have a strong attenuation. Neutron radiography can perform many detection tasks that X-ray and gamma-ray radiography cannot. Neutron radiography is also a very efficient technique which is suitable to investigate the transport and distribution of water in wood, and the spatial resolution of the neutron radiographs is 85-100 μm (Islam et al. 2003). For example, Mannes *et al.* demonstrated that even small amounts of water absorbed from air moisture can be detected and quantified by this technique (Mannes et al. 2009). Sedighi-Gilani *et al.* revealed the interesting features about the process of liquid uptake and quantification and visualization of absorbed water, with a resolution of 0.07 kg/m^3 by the neutron radiographs (Sedighi-Gilani et al. 2012). Non-destructive neutron radiography is used to study the different processes of liquid transport in Scots pine sapwood and heartwood. The spatial and temporal changes in moisture content and saturation degree, measured at high resolution, are provided for water uptake in longitudinal, radial and tangential directions.

Neutron radiography on the other hand has been successful in quantifying the water transport and visualization of spatial distribution of water content in porous media with a high resolution, up to 0.1 kg/m^3. This accuracy is due to the high attenuation of the neutron beam by hydrogen nuclei, a component of water, which provides a high contrast for water versus the other wood components. Although there are many advantages, but there are disadvantages, harmful to human health, it will produce radiation safety problems.

1.2.4 Vibrational spectroscopic imaging techniques

Vibrational spectroscopy can be used to measure the vibrational spectra of samples at the micron or even submicron scale. It can not only obtain the molecular structure information but also the spatial distribution information of the chemical components, but also keep the characteristics of vibrational spectra such as labelessness, directness, nondestructiveness, simplicity and rapidity. It has been used in many fields. More recently, vibrational spectroscopic imaging techniques such as micro-FTIR and confocal Raman spectrometer equipped with additional visible-light microscope could offer visual examination and spectral information of chemical functional groups *in situ*, and these techniques yield high spatial resolutions at the micro scale. So these two techniques could provide *in situ* and high-resolution observation of water distribution in wood that can be complementary to the techniques discussed above.

These two vibrational spectroscopic imaging techniques play complementary roles in the physical selection rules and spatial resolutions. For the physical selection rules, the peaks of micro-FTIR are caused by the change of electric dipole moment of sample molecules, and the peaks of Raman are determined by electric dipole-electric dipole polarizability. Hence, there is a comprehensive study on adsorbed water for vibrational spectroscopic imaging techniques. For the spatial resolutions, the lateral resolution of micro-FTIR spectroscopy is 6.25 μm by the instrumental parameters, and the lateral resolution of confocal Raman spectroscopy is 1μm in this paper. Due to the limited resolution, μ-FTIR spectroscopy can be used to observe the adsorbed water in different wood cell walls and confocal Raman spectroscopy is appropriate for investigating adsorbed water in morphologically distinct regions of the wood cell wall. For example ,Guo *et al.* used micro-FTIR to detect different areas of cell wall of Ginkgo biloba L., and based on the vibration of OH, C=O and C—O groups, detected the adsorption position of water. At the same time, confocal Raman spectroscopy was used to collect image from cell corner (CC) and middle layer of secondary wall (S2) and non-uniform wall, which also confirmed the distribution of water. Although the spatial resolution of vibrational spectral imaging is very high, the confocal Raman is limited, the sample must be thin, and there is only one point in the Raman mode (Guo et al. 2017).

1.3 Determining molecular interactions between adsorbed water and wood

One of the most important factors affecting the water adsorption of wood is molecular interaction occurring between the wood and adsorbed water. The first main purpose has been to gather information on the water sorption sites. Meanwhile, many studies have led to the identification of several different types of water adsorbed in wood. Generally, there are two types of water in wood. The "water of constitution" is the water included in the chemical structure of wood and it is inherent to the organic nature of the wood cell walls. It cannot be removed without modifying the chemical composition of wood. The second type of water comes in three forms: "bound" or "hygroscopic" water which is adsorbed by sorption sites in amorphous areas of cellulose and hemicelluloses present in the cell walls; "free" water which is liquid like water in the cell lumen and voids; and "vapor" water which is present in voids and cell lumen unless the wood is completely saturated. Many methods such as NIR spectroscopy, Nuclear Magnetic Resonance, FTIR spectroscopy and Raman spectroscopy have been widely employed to investigate molecular association of adsorbed water with wood.

1.3.1 Near-infrared spectroscopy

Near-infrared spectroscopy is mainly caused by the non-resonant nature of molecular vibrations, which makes the molecular vibrations transition from the ground state to the high energy level. Near-infrared spectroscopy can not only reflect the composition and structural properties of most organic compounds, but also indirectly reflect the existing information of some inorganic ion compounds through the spectral changes caused by its influence on the coexisting substance. NIR spectroscopy has many advantages, such as high speed, high efficiency, low cost, good reproducibility and easy measurement, so it plays an important role in wood science research and production.

The NIR technique enables overtones and combination bands from molecular vibrations in the IR region to be studied and is thus assumed to be sensitive enough to distinguish between water molecules adsorbed onto different hydrophilic groups (Tsuchikawa et al. 2003). The NIR portion of the electromagnetic spectrum contains information on more than one hydrogen-bonded species at the water bands at 1450 nm and 1930 nm. These bands consist of multiple overlapping peaks, and it is known that variations in the hydrogen-bonded subspecies cause band broadening and peak position shifts. Using NIR combined with multivariate statistical analysis, it has been demonstrated that it is possible to obtain chemical information related to the adsorption of water in CMC systems. Furthermore, it is shown that it is possible to distinguish between two kinds of adsorbed water; non-freezing bound water adsorbed by carboxyl acid groups and by hydroxyl groups.

1.3.2 Nuclear magnetic resonance technique

Nuclear magnetic resonance is a phenomenon in which the nucleus is excited by external RF, absorbs the energy carried by RF pulses, and reaches a new equilibrium level transition. When the RF pulse is removed, the nucleus recovers to the lower energy level according to the principle of minimum energy. This process releases the absorbed energy and generates the NMR signal. Many physical and chemical information of substances can be obtained by collecting and analyzing these MRI signals in practical applications. The NMR spectra are formed by the absorption of polar molecules or water molecules in wood materials, and the structure, defects and valuable information of wood materials are analyzed. There are many work reported on the use of NMR to study wood-water interactions (Brownstein 1980; Hall and Rajanayagam 1986; Menon *et al*. 1987; Riggin et al. 1979). For example, Froix and Nelson measured both T1 and T2 relaxation times for cotton linters in a range of 0-25% moisture content (Froix and Nelson 1975). Four different states of water were identified: primary bound water on the cellulose crystal, and two types of secondary bound water associated with the cellulose structure as well as bulk water. Menon *et al*. did a comprehensive study on water in wood, finding three pools of water which they assigned to the cell wall, the ray and tracheid lumens and the earlywood tracheid lumens (Menon et al. 1987). Also by choosing different tree species, they found that species had an effect on the T2 values. Species with smaller cell lumens had shorter T2 values for the lumen water. Araujo *et al.* examined the location of water in white spruce softwood by NMR identifying bound water, lumen water in late wood cells and lumen water in early wood cells. The latter having the largest lumens and thus the longest relaxation time (Araujo et al. 1992).

By NMR, the amount of water (free or bound water) is measured directly, although in practice a simple calibration needs to be carried out. The only drawback of nuclear magnetic resonance is that the equipment is too expensive for production.

1.3.3 Fourier transform infrared spectroscopy

In recent years, the water problem of wood has become more and more serious, and the molecular interaction between wood and adsorbed water has become the focus of research. Among many methods, FTIR has "fingerprint technique", which means that any two samples, even if only slightly different, will never have the same spectrum, which is mainly reflected in the number of peaks, peak migration, the intensity of the maximum peak, etc. This method has the advantages of high sensitivity, good reproducibility, high speed, no need of sample pretreatment, simultaneous determination of multi-components and on-line detection. FTIR spectroscopy has been widely used to elucidate the chemical structure of wood in its natural state (Bouhacina et al. 2000; Ping et al. 2001). For example, Evans proved that it was possible to differentiate between hardwoods and softwoods by FTIR spectroscopy (Evans 1991). Guo et al. showed the degradation of xylan and glucomannan, and the loss of $C=O$ groups linked to the aromatic skeleton in lignin during mechanical compression combined with steam treatment on spruce wood by FTIR spectroscopy (Guo et al. 2015). Meanwhile, FTIR spectroscopy has been exploited previously to study the moisture adsorption in cellulosic materials. Hofstetter et al. investigated the interaction between cellulose and moisture by the combination of deuteration and FTIR spectroscopy, and showed the role of different hydrogen bonds in the moisture uptake (Hofstetter et al. 2006). Olsson et al. examined the adsorption characteristics of some wood polymers by FTIR spectroscopy in humid atmospheres at different RH levels, and determined one important relationship between the moisture weight gain and the increase of the $O-H$ stretching envelope (Olsson and Salmen 2004). Laity et al. demonstrated that it was possible to reproduce water sorption kinetics by recording FTIR spectra (Laity and Hay 2000). Célino et al. showed that the spectral information of FTIR spectra allowed both qualitative and quantitative analyses of the moisture absorption mechanisms of natural fibers (Célino et al. 2014). Meanwhile, better sensitivity in the detection of the water content can be achieved using micro-FTIR spectroscopy. Liu et al. showed that micro-FTIR spectroscopy was relatively simple, inexpensive, and available for quantitatively investigations of the water adsorption of micron-sized samples (Liu et al. 2008). Furthermore, the idea that moisture is adsorbed to specific sites on the lignocellulosic material has also been advocated by FTIR spectroscopy. Haxaire et al. showed that infrared spectroscopy can help to identify whether the hydroxyl groups are hydrogen bonded or not with water molecules (Haxaire et al. 2003). Olsson et al. examined the association of water on pulp paper using FTIR spectra, and indicated all the moisture-sorbing sites adsorbed moisture to the same relative degree (Olsson and Salmen 2004). A limited number of 1.0-1.3 adsorbed water molecules per hydroxyl group at relative humidities below 100% had been established for a number of hydrophilic materials. Cellulose is the most important component in terms of its volume and its effect on the characteristics of wood. For a model system of partly carboxymethylated cellulose, the specificity of moisture adsorption to hydroxyl and carboxyl groups was demonstrated (Berthold et al. 1998; Kachrimanis et al. 2006). At the same time, Fourier transform infrared instrument is simple and cheap, which is suitable for the study of wood adsorbed water.

1.3.4 Raman spectroscopy

Raman spectroscopy is a technique developed in 1920s. Raman spectroscopy has been paid much attention by researchers for its advantages of non-destructiveness, rich information and no need of sample preparation. Meanwhile, Raman spectroscopy is also a powerful tool for analyzing wood without the need of using a stain, or a dye, or a contrast agent for labeling (Gierlinger and Schwanninger 2006; Guo et al. 2010; Sun et al. 2011). It has been applied to detect the distribution of extractives of wood, in situ analyze the chemical groups of wood, observe the distribution of lignin and cellulose, and investigate tension wood fibers and so on (Hanninen et al. 2011; Ma et al. 2013). For example, Agarwal examined the lignin and cellulose distribution in black spruce latewood cell walls and indicated that the concentrations of both lignin and cellulose varied both within and between distinct morphological regions, this conclusion was supported by other studies (Roder et al. 2004; Zhang et al. 2012). Gierlinger *et al.* investigated one single wood fiber by Raman spectra, and showed that the band in the OH-stretching region could provide information about the strength of hydrogen bonding (Gierlinger et al. 2006). Furthermore, Raman spectroscopy has been exploited previously to study the moisture adsorption or evaporation in cellulosic and lignocellulosic materials. Agarwal *et al.* demonstrated that the water evaporation in the cellulose filter paper and spruce thermomechanical pulp has a linear dependence relation with the declining Raman intensity of the O—H stretching envelope (Agarwal and Kawai 2005). Scherer *et al.* observed cellulose acetate film which was exposed to water vapor from 0% to 100% relative humidity, and characterized the water adsorption by the increasing O—H stretching envelope as a function of RH (Scherer et al. 1985). Meanwhile, due to the high spatial resolution of confocal Raman spectroscopy, it has been successfully used to study the water adsorbed by micron-sized samples. Recently, Raman spectroscopy has been developed to determine the current state of water in wood. However, this approach also has some limitations. Errors may occur due to the fluorescence of the wood.

1.4 Future directions

Although the results are promising, all these mentioned techniques still suffer from resolution-related issues, the size of the measured area, measurable moisture content range level, the size of the sample, and measurement time. Therefore, new approaches with greater accuracy and higher resolution are encouraged to apply in the research of wood-water relationships.

Meanwhile, the wood cell wall contains cellulose, hemicelluloses, and lignin which have different physical and chemical properties and bioavailability. And the hydration structures of these chemical compositions are still unclear. Because of the different possibilities of hydrogen-bond formation in adsorption sites, the adsorbed water exists in a special form deviated from ordinary water and the interaction between wood and water is very complicated, which deserves a more detailed investigation.

Furthermore, there exists limited knowledge about water adsorption behavior among different morphological regions of wood cell walls. A better understanding of water adsorption of wood cell walls at the micro scale is urgently needed for developing more accurate kinetic models for moisture adsorption and moisture movement in wood.

References

AGARWAL U P. 2006. Raman imaging to investigate ultrastructure and composition of plant cell walls: Distribution of lignin and cellulose in black spruce wood (Picea mariana) [J]. Planta, 224 (5): 1141-1153.

AGARWAL U P, KAWAI N. 2005. "Self-absorption" phenomenon in near-infrared Fourier transform Raman spectroscopy of cellulosic and lignocellulosic materials [J]. Applied Spectroscopy, 59 (3): 385-388.

ARAUJO C D, MACKAY A L, HAILEY J, et al. 1992. Proton magnetic-resonance techniques for characterization of water in wood - application to white spruce [J]. Wood Science and Technology, 26 (2): 101-113.

BAYER E A, SHOHAM Y, LAMED R. 2006. Cellulose-decomposing bacteria and their enzyme systems [J]. The prokaryotes, 2 (Part 1): 578-617.

BERTHOLD J, OLSSON R, SALMEN L. 1998. Water sorption to hydroxyl and carboxylic acid groups in carboxymethylcellulose (CMC) studied with NIR-spectroscopy [J]. Cellulose, 5 (4):

281-298.

BOUHACINA T, DESBAT B, AIME J P. 2000. FTIR spectroscopy and nanotribological comparative studies: influence of the adsorbed water layers on the tribological behavior [J]. Tribology Letters, 9 (1-2): 111-117.

BRANDSTROM J. 2001. Micro- and ultrastructural aspects of Norway spruce tracheids: A review [J]. Iawa Journal, 22 (4): 333-353.

BROWNSTEIN K R. 1980. Diffusion as an explanation of observed NMR behavior of water absorbed on wood [J]. Journal of Magnetic Resonance, 40 (3): 505-510.

CASIERI C, SENNI L, ROMAGNOLI M, et al. 2004. Determination of moisture fraction in wood by mobile NMR device [J]. Journal of Magnetic Resonance, 171 (2): 364-372.

CÉLINO A, GONCALVES O, JACQUEMIN F, et al. 2014. Qualitative and quantitative assessment of water sorption in natural fibers using ATR-FT'IR spectroscopy [J]. Carbohydrate Polymers, 101: 163-170.

CHAUHAN S S, AGGARWAL P. 2004. Effect of moisture sorption state on transverse dimensional changes in wood [J]. Holz Als Roh-Und Werkstoff, 62 (1): 50-55.

DAVIS J R, ILIC J, WELLS P. 1993. Moisture-content in drying wood using direct scanning gamma-ray densitometry [J]. Wood and Fiber Science, 25 (2): 153-162.

DVINSKIKH S V, HENRIKSSON M, BERGLUND L A, et al. 2011. A multinuclear magnetic resonance imaging (MRI) study of wood with adsorbed water: Estimating bound water concentration and local wood density [J]. Holzforschung, 65 (1): 103-107.

EVANS P A. 1991. Differentiating hard from soft woods using Fourier-transform infrared and Fourier-transform Raman spectroscopy [J]. Spectrochimica Acta Part A-Molecular and Biomolecular Spectroscopy, 47 (9-10): 1441-1447.

EVANS P D, BANKS W B. 1988. Degradation of wood surfaces by water-changes in mechanical-properties of thin wood strips [J]. Holz Als Roh-Und Werkstoff, 46 (11): 427-435.

FROIX M F, NELSON R. 1975. The interaction of water with cellulose from nuclear magnetic resonance relaxation times [J]. Macromolecules, 8 (6): 726-730.

FROMM J H, SAUTTER I, MATTHIES D, et al. 2001. Xylem water content and wood density in spruce and oak trees detected by high-resolution computed tomography [J]. Plant Physiology, 127 (2): 416-425.

GIERLINGER N, SCHWANNINGER M. 2006. Chemical imaging of poplar wood cell walls by confocal Raman microscopy [J]. Plant Physiology, 140 (4): 1246-1254.

GIERLINGER N, SCHWANNINGER M, REINECKE A, et al. 2006. Molecular changes during tensile deformation of single wood fibers followed by Raman microscopy [J]. Biomacromolecules, 7 (7): 2077-2081.

GUINDOS P. 2014. Numerical modeling of timber with knots: the progressively damaged lattice approach vs. the equivalent damaged continuum [J]. Holzforschung, 68 (5): 599-613.

GUO J, SONG K, SALMEN L, et al. 2015. Changes of wood cell walls in response to hygro-mechanical steam treatment [J]. Carbohydrate Polymers, 115: 207-214.

GUO X, SHOU J, ZHANG Y, et al. 2010. Micro-Raman analysis of association equilibria in supersaturated $NaClO_4$ droplets [J]. Analyst, 135 (3): 495-502.

GUO X, WU Y, YAN N. 2017. Characterizing spatial distribution of the adsorbed water in wood cell wall of Ginkgo biloba L. by μ-FTIR and confocal Raman spectroscopy [J]. Holzforschung, 71 (5): 415-423.

HALL L D, RAJANAYAGAM V. 1986. Evaluation of the distribution of water in wood by use of 3-D proton NMR volume imaging [J]. Wood Science and Technology, 20 (4): 329-333.

HAMAD W Y, GURNAGUL N, GULATI D. 2012. Analysis of fibre deformation processes in high-consistency refining based on Raman microscopy and X-ray diffraction [J]. Holzforschung, 66 (6): 711-716.

HANNINEN T, KONTTURI E, VUORINEN T. 2011. Distribution of lignin and its coniferyl alcohol and coniferyl aldehyde groups in Picea abies and Pinus sylvestris as observed by Raman imaging [J]. Phytochemistry, 72 (14-15): 1889-1895.

HATFIELD R, VERMERRIS W. 2001. Lignin formation in plants. The dilemma of linkage specificity [J]. Plant Physiology, 126 (4): 1351-1357.

HAXAIRE K, MARECHAL Y, MILAS M, et al. 2003. Hydration of polysaccharide hyaluronan observed by IR spectrometry. I. Preliminary experiments and band assignments [J]. Biopolymers, 72 (1): 10-20.

HOFSTETTER K, HINTERSTOISSER B, SALMEN L. 2006. Moisture uptake in native cellulose-the roles of different hydrogen bonds: a dynamic FTIR study using deuterium exchange [J]. Cellulose, 13 (2): 131-145.

HOZJAN T, SVENSSON S. 2011. Theoretical analysis of moisture transport in wood as an open porous hygroscopic material [J]. Holzforschung, 65 (1): 97-102.

ISHIMARU Y, ARAI K, MIZUTANI M, et al. 2001. Physical and mechanical properties of wood after moisture conditioning [J]. Journal of Wood Science, 47 (3): 185-191.

ISLAM M N, KHAN M A, ALAM M K, et al. 2003. Study of water absorption behavior in wood plastic composites by using neutron radiography techniques [J]. Polymer-Plastics Technology and Engineering, 42 (5): 925-934.

KACHRIMANIS K, NOISTERNIG M F, GRIESSER U J, et al. 2006. Dynamic moisture sorption and desorption of standard and silicified microcrystalline cellulose [J]. European Journal of Pharmaceutics and Biopharmaceutics, 64 (3): 307-315.

KARAASLAN A M, TSHABALALA M A, BUSCHLE-DILLER G. 2010. Wood hemicellulose/ chitosan-based semi-interpenetrating network hydrogels: mechanical, swelling and controlled drug release properties [J]. BioResources, 5 (2): 1036-1054.

KHALIL H P S A, ALWANI M S, OMAR A K M. 2007. Cell wall structure of various types of tropical plant waste fibers [J]. Journal of the Korean Wood Science and Technology, 35 (2): 9-15.

LAITY P R, HAY J N. 2000. Measurement of water diffusion through cellophane using attenuated total reflectance-Fourier transform infrared spectroscopy [J]. Cellulose, 7 (4): 387-397.

LEE W Y. 1981. Ultrastructure of wood cell wall tracheids: the structure of spiral thickenings in compression wood [J]. Wood Science & Technology, 9 (1): 1-12.

LI S, ZHANG L, WANG Y, et al. 2017. Knockdown of a cellulose synthase gene BoiCesA affects the leaf anatomy, cellulose content and salt tolerance in broccoli [J]. Scientific Reports, 7 (1): 1-14.

LISPERGUER J, PEREZ P, URIZAR S. 2009. Structure and thermal properties of lignins: characterization by infrared spectroscopy and differential scanning calorimetry [J]. Journal of the Chilean Chemical Society, 54 (4): 460-463.

LIU Y, YANG Z, DESYATERIK Y, et al. 2008. Hygroscopic behavior of substrate-deposited particles studied by micro-FTIR spectroscopy and complementary methods of particle analysis [J]. Analytical Chemistry, 80 (3): 633-642.

MA J, JI Z, ZHOU X, et al. 2013. Transmission electron microscopy, fluorescence microscopy, and confocal raman microscopic analysis of ultrastructural and compositional heterogeneity of cornus alba l. wood cell wall [J]. Microscopy and Microanalysis, 19 (1): 243-253.

MANNES D, SONDEREGGER W, HERING S, et al. 2009. Non-destructive determination and quantification of diffusion processes in wood by means of neutron imaging [J]. Holzforschung, 63 (5): 589-596.

MELLEROWICZ E J, SUNDBERG B. 2008. Wood cell walls: biosynthesis, developmental dynamics and their implications for wood properties [J]. Current Opinion in Plant Biology, 11 (3): 293-300.

MENON R S, MACKAY A L, HAILEY J, et al. 1987. An NMR determination of the physiological water distribution in wood during drying [J]. Journal of Applied Polymer Science, 33 (4): 1141-1155.

MERELA M, SEPE A, OVEN P, et al. 2005. Three-dimensional *in vivo* magnetic resonance microscopy of beech (Fagus sylvatica L.) wood [J]. Magnetic Resonance Materials in Physics Biology and Medicine, 18 (4): 171-174.

MOLINSKI W, RACZKOWSKI J. 1988. Mechanical stresses generated by water adsorption in wood and their determination by tension creep measurements [J]. Wood Science and Technology, 22 (3): 193-198.

OLEK W, MAJKA J, CZAJKOWSKI L. 2013. Sorption isotherms of thermally modified wood [J]. Holzforschung, 67 (2): 183-191.

OLSSON A M, SALMEN L. 2004. The association of water to cellulose and hemicellulose in paper examined by FTIR spectroscopy [J]. Carbohydrate Research, 339 (4): 813-818.

PASSARINI L, MALVEAU C, HERNANDEZ R E. 2015. Distribution of the equilibrium moisture content in four hardwoods below fiber saturation point with magnetic resonance microimaging [J]. Wood Science and Technology, 49 (6): 1251-1268.

PENG L, ZHANG L, CHENG X, et al. 2013. Disruption of cellulose synthesis by 2,6-dichlorobenzonitrile affects the structure of the cytoskeleton and cell wall construction in Arabidopsis [J]. Plant Biology, 15 (2): 405-414.

PING Z H, NGUYEN Q T, CHEN S M, et al. 2001. States of water in different hydrophilic polymers-DSC and FTIR studies [J]. Polymer, 42 (20): 8461-8467.

RALPH J, LUNDQUIST K, BRUNOW G, et al. 2004. Lignins: Natural polymers from oxidative coupling of 4-hydroxyphenyl-propanoids [J]. Phytochemistry Reviews, 3 (1-2): 29-60.

RIGGIN M T, SHARP A R, KAISER R, et al. 1979. Transverse NMR relaxation of water in wood [J]. Journal of Applied Polymer Science, 23 (11): 3147-3154.

RODER T, KOCH G, SIXTA H. 2004. Application of confocal Raman spectroscopy for the topochemical distribution of lignin and cellulose in plant cell walls of beech wood (Fagus sylvatica L.) compared to UV microspectrophotometry [J]. Holzforschung, 58 (5): 480-482.

ROSENKILDE A, GLOVER P. 2002. High resolution measurement of the surface layer moisture content during drying of wood using a novel magnetic resonance imaging technique [J]. Holzforschung, 56 (3): 312-317.

RUIZ-DUENAS F J, MARTINEZ A T. 2009. Microbial degradation of lignin: How a bulky recalcitrant polymer is efficiently recycled in nature and how we can take advantage of this [J]. Microbial Biotechnology, 2 (2SI): 164-177.

SAHA B C. 2003. Hemicellulose bioconversion [J]. Journal of Industrial Microbiology & Biotechnology, 30 (5): 279-291.

SAKAGAMI H, HASHIMOTO K, SUZUKI F, et al. 2005. Molecular requirements of lignin-carbohydrate complexes for expression of unique biological activities [J]. Phytochemistry, 66 (17): 2108-2120.

SCHERER J R, BAILEY G F, KINT S, et al. 1985. Water in polymer membranes. 4. Raman-scattering from cellulose-acetate films [J]. Journal of Physical Chemistry, 89 (2): 312-319.

SEDIGHI-GILANI M, GRIFFA M, MANNES D, et al. 2012. Visualization and quantification of liquid water transport in softwood by means of neutron radiography [J]. International Journal of Heat and Mass Transfer, 55 (21-22): 6211-6221.

SUN L, SIMMONS B A, SINGH S. 2011. Understanding tissue specific compositions of bioenergy feedstocks through hyperspectral Raman imaging [J]. Biotechnology and Bioengineering, 108 (2): 286-295.

TERASHIMA N, KITANO K, KOJIMA M, et al. 2009. Nanostructural assembly of cellulose, hemicellulose, and lignin in the middle layer of secondary wall of ginkgo tracheid [J]. Journal of Wood Science, 55 (6): 409-416.

TSUCHIKAWA S, INOUE K, NOMA J, et al. 2003. Application of near-infrared spectroscopy to wood discrimination [J]. Journal of Wood Science, 49 (1): 29-35.

TSUCHIKAWA S, SIESLER H W. 2003. Near-infrared spectroscopic monitoring of the diffusion process of deuterium-labeled molecules in wood. Part I: softwood [J]. Applied Spectroscopy, 57 (6): 667-674.

YONG S Y C, WICKNESWARI R. 2013. Molecular characterization of a cellulose synthase gene (AaxmCesA1) isolated from an Acacia auriculiformis x Acacia mangium hybrid [J]. Plant Molecular Biology Reporter, 31 (2): 303-313.

ZHANG H, YANG H, GUO H, et al. 2014. Kinetic study on the liquefaction of wood and its three cell wall component in polyhydric alcohols [J]. Applied Energy, 113 (SI): 1596-1600.

ZHANG Z H, JI Z, MA J F, et al. 2012. Anatomy, cell wall ultrastructure and inhomogeneity in lignin distribution of broussonetia papyrifera [J]. Cellulose Chemistry and Technology, 46 (3-4): 157-164.

Chapter 2 Quantitative detection of moisture content in heat-treated wood cell walls using micro-FTIR spectroscopy

2.1 Introduction

Wood, as the most important renewable resource in the world, with its unique carbon fixation, green, easy processing, renewable and other performance characteristics, occupies an increasingly important position in life, production and other aspects. It has a variety of uses, which can be widely used in construction, paper, textile, and applied chemical industries. However, wood is a hygroscopic material (Esteban et al. 2006; Olek et al. 2013; Zhang et al. 2016). When the wood is placed in a humid environment, there is an exchange of energy and substances between the moisture in the air and the wood. The relationship between them is extremely complex and has always been the focus and difficulty of the field of wood science research. This complex relationship between wood and moisture not only exists in the formation of wood, but also plays an important role in its manufacture, processing, and use. Below the fiber saturation point, moisture has the most significant effect on wood, and its dimensions and density (Engelund et al. 2013; Searson et al. 2004), as well as its mechanical (Evans and Banks 1988; Moliński and Raczkowski 1988; Ouertani et al. 2014), elastic (Hogan Jr and Niklas 2004; Irvine and Grace 1997; Maeda and Fukada 1987), and electrical properties (Zelinka et al. 2008) are affected by its moisture content. In order to improve the hygroscopicity of wood, many methods have been advocated, in which heat modification has been proven to significantly reduce water adsorption (Borrega and Karenlampi 2010; Huang et al. 2012; Kartal et al. 2007; Scheiding et al. 2016; Srinivas and Pandey 2012). In addition, the heat treatment modification technology does not use any chemical agents in the treatment process, and the heat treatment wood will not release toxic and harmful substances during the use process, which belongs to an energy-saving and environmentally friendly green treatment method. At present, the moisture absorption of heat-treated wood has received widespread attention. A fundamental understanding of water adsorption in heat-treated wood is very important with respect to heat-treated wood's commercial utilization as well as its biology and chemistry.

As reported in the literature, one of the original intentions of wood heat treatment is to reduce moisture absorption. In general, when wood is treated at 180 ℃ or higher, chemical changes such as hemicellulose degradation can cause a decrease in hygroscopicity. Heat treatment reduces the relative content of polysaccharides in wood, resulting in a reduction in the number of carbonyl groups, free hydroxyl groups and other highly hygroscopic groups. In addition, the cross-linking of lignin during heat treatment may also reduce the hygroscopicity of wood, because the cross-linking reaction reduces the volume of water that can be adsorbed between adjacent polymers of the cell wall. Regarding the hygroscopicity of heat-treated wood, a large amount of literature has been reported in recent years (Awoyemi et al. 2009; Dilik and Hiziroglu 2012; Joma et al. 2017; Salca and Hiziroglu 2014; Wang et al. 2011; Wiberg and Morén 1999; Willems et al. 2015). Almeida *et al.* (Almeida et al. 2009) studied wood-water relationship of untreated and heat-treated wood, and then determined the following parameters, such as fiber saturation point, wood anisotropy, shrinkage slope, reduction in hygroscopicity, and anti-shrink efficiency. Metsä-Kortelainen *et al.* (Metsä-Kortelainen et al. 2006) examined the hygroscopic property of Scots pine (Pinus sylvestris) and Norway spruce (Picea abies) and found that heat-treatment evidently decreased the water absorption of these two kinds of wood. Besides, Hill *et al.* (Hill et al. 2012) determined the water vapor sorption behavior of heat-treated Scots pine (Pinus sylvestris L.) using dynamic vapor sorption (DVS) apparatus and analyzed the resulting kinetic curves by the parallel exponential kinetics model. Further, the sorption isotherms and sorption hysteresis of other heat-treated wood including acacia (Acacia mangium) (Willems 2014), sesendok (Endospermum malaccense) (Jalaludin et al. 2010), and eucalyptus pellita (Sun et al. 2017) were all investigated using DVS method. It is generally accepted that DVS method can provide vast highly reproducible water sorption data (Hill et al. 2010, 2010), so this method can be used to offer reference value for water adsorption in the heat-treated wood.

With the deep research of water adsorption in the heat-treated wood, some spectroscopic techniques have been exploited, including UV-visible spectroscopy, near-infrared spectroscopy, Fourier transform infrared spectroscopy, Raman spectroscopy, mass spectrometry, nuclear magnetic resonance spectroscopy, etc. Among these methods, the near-infrared spectroscopy (Ferraz et al. 2005; Inagaki et al. 2008; Lestander 2008; Mitsui et al. 2008; Pentrak et al. 2012; Popescu et al. 2016; Sandak et al. 2015), Fourier transform infrared spectroscopy (FTIR) (Akgül et al. 2007; Guo et al. 2017; Kotilainen et al. 2000; Ozgenc et al. 2017), nuclear magnetic resonance (NMR) spectroscopy (Bardet et al. 2004; Casieri et al. 2004; Hsi et al. 1977; Kekkonen et al. 2014; Senni et al. 2009; Turov and Leboda 1999; Zhang 2011) and Raman spectroscopy (Atalla 1987; Ding et al. 2016) have been used to study the water sorption. Esteves *et al.* (Esteves and Pereira 2008) used near-infrared spectroscopy to predict moisture content of heat-treated pine (Pinus pinaster) and eucalypt (Eucalyptus globulus), and revealed the coefficients of determination range from 78% to 95%. Boonstra *et al.* (Boonstra and Tjeerdsma 2006) used FTIR spectroscopy to study water adsorption of heat-treated wood, and revealed that the reduction of the hygroscopicity was due to the cross-linking of the lignin and the decrease of OH groups. Ding *et al.* (Ding et al. 2016) used Raman spectroscopy to analyze the water sorption in the heat-treated pine wood, and confirmed that lignin was one of the key factors affecting the sorption behavior of the heat-treated wood. Among these spectroscopic techniques, FTIR spectroscopy has been widely applied, for it has unique advantages, i.e., the easily distinguishable characteristic peak (Belfer et al. 2000; Benar et al. 1999; Colom and Carrillo 2005; Jiang et al. 2016; Ling et al. 2013; Oh et al. 2005), the high sensitivity (Beskers et al. 2015; Jackson et al. 1993), and the good accuracy of spectral analysis (Pandey et al. 2012; Samuel and Mohan 2004). Moreover, the micro-FTIR spectroscopy can combine the visual imaging of microscope with the chemical analysis of functional groups in the infrared spectrum, which can not only image the topography of the object, but also provide the spectral information of each point in the object space. Because it has the advantages of non-destructive testing, micro area analysis, extensive analysis of samples and spatial distribution of functional groups, the development of micro-FTIR spectroscopy is convenient for detecting water adsorption (Guo and Wu 2018; Ito and Nakashima 2002).

As micro-FTIR spectroscopy has the largest potential, the aim of this study is to explore the possibility of this technique for quantitative assessment of moisture sorption in nanogram quantities of heat-treated wood. Firstly, we measured micro-FTIR spectra of heat-treated wood at different RH levels. Secondly, we obtained MC of heat-treated wood at the same RH levels by DVS apparatus which were applied as reference values. Thirdly, taking advantage of the identified spectral regions of micro-FTIR spectra and collected reference values, we built and developed multivariate models using PLS-R method. Finally, we used the optimal multivariate model to estimate sorption isotherm, and demonstrated the validity of micro-FTIR spectra in quantitative detection of water adsorption in heat-treated wood.

2.2 Materials and methods

2.2.1 Materials

Three wood specimens (dimensions 100 mm × 30 mm × 20 mm in length, width, and thickness) were cut from straight stem of Ginkgo biloba L. (Ginkgoaceae). Then heat treatment was used for these wood specimens in electric vacuum drying oven under controlled condition of (180±1)℃. This heat treatment lasted 4 h. From these heat-treated wood specimens, transverse sections were prepared without embedding and any chemical treatment. These sections were cut using a manual rotary microtome (Leica RM2135), and then placed on the bottom of sample cell. Prior to the spectral measurement, wood transverse section was air-dried for 2h.

2.2.2 Micro-FTIR spectrometer

The experimental apparatus is presented in Figure 2.1 (a). The main section was a spectrometer (Nicolet IN 10TM), which was applied for recording micro-FTIR spectral development of wood *vs*. RH. This main section was installed with one microscope which provided new function of visual examination and selecting observation area. In spectral measurement, all micro-FTIR spectra were acquired from one randomly selected area (50 μm × 50 μm), in which only wood cell wall was present (~1 ng). These spectra in the range of 720-4000 cm^{-1} were accumulated 32 times by scanning the grating at the spectral resolution of 4 cm^{-1}.

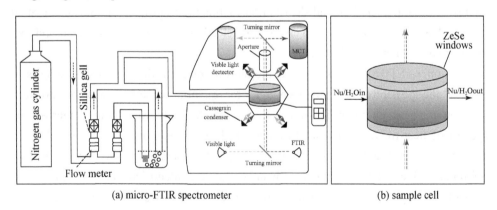

(a) micro-FTIR spectrometer   (b) sample cell

Figure 2.1   Experimental setup for measurement of micro-FTIR spectroscopy

Figure 2.1 (b) also displays the schematic diagram of sample cell. First, one wood transverse section was mounted onto the bottom of sample cell. Then, this cell was sealed, and placed on the automatic stage of micro-FTIR spectrometer. Through the cell, Nitrogen gas with specific RH was circulated. Spectral measurements were carried out during the RH range from 0% to 90% at the constant temperature of 25 ℃. Representative variations of target RH, actual RH, and recorded micro-FTIR spectra over time were displayed in Figure 2.2. When target RH was set the next value (e.g., 5% and 10% RH), the time delay of 3-4 min would appear. During this time, actual RH got close to the target RH, and then remained stable throughout. After 15 min, the spectra collected every 0.5 min remained the same up to 200 min (main peak height of the 3358 cm^{-1} band was advocated to detect spectral change). Based on these results, 60 min were set as balance time at each RH level.

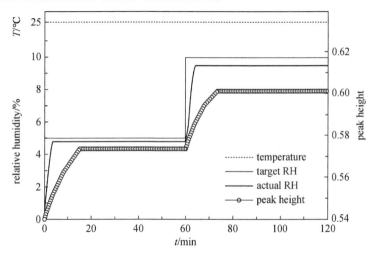

Figure 2.2　Typical changes of the set RH, the real RH, temperature and peak height of real-time spectrum with time

2.2.3　DVS apparatus

Sorption isotherm was measured using the DVS equipment (DVS AdvantagePlus, Surface Measurement Systems Ltd, London, United Kingdom). First, heat-treated wood specimen was put on a metal plate. Then this plate was connected to the microbalance which was situated in the temperature and humidity auto-controlled box. A typical run started at 0% RH and then increased in RH steps in 5% steps from 0% to 95%, before decreasing to 0% RH in the same step. In this process, the certain constant temperature of 25 ℃ was kept. Meanwhile, every set RH was maintained for enough time until sample mass altered less than 0.002% per minute during 10 min.

In the sorption isotherm, the moisture content was calculated based on the recorded sample mass and the following formula:

$$MC = \frac{m - m_d}{m_d} \times 100\% \quad (2.1)$$

Where, MC was moisture content, m_d was dry mass, and m was the recorded sample mass.

Typical changes of MC and RH vs. time were displayed in Figure 2.3. When target RH was set the next value (e.g., 5% and 10% RH), the time delay of 4-10 min would appear. During this time, actual RH got close to target RH, and then remained stable throughout. Meanwhile, the MC followed the asymptotic curve, and then achieved balance when was recorded as reference value. Further, the temperature values were very stable during this process. It should be noticed that three replicates of these heat-treated wood specimens were exposed and each MC collected as reference value in the sorption isotherm was mean of three replicates.

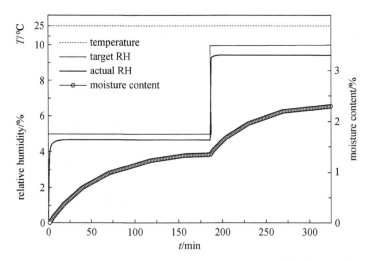

Figure 2.3 Typical changes of MC and RH vs. time

2.2.4 Micro-FTIR data processing

1. Acquirement of difference spectrum

To gain the detailed spectral change of heat-treated wood, FTIR difference spectrum technique was employed in the OMINIC 8.0 (Thermo Scientific Inc., Madison, Wisconsin, USA). By taking advantage of this technique, difference spectrum was obtained by subtracting the micro-FTIR spectrum collected at the RH of 0%.

2. Establishment of multivariate quantitative model

Multivariate quantitative model could be generated using PLSR approach. The pathlength type could be set to "constant". In components table, the only component, MC was list, and maximum and minimum values of MC were set to "17.3%" and "0%" separately, which were from the reference values collected by DVS apparatus. Then original micro-FTIR spectra of wood collected at sixteen levels of RH were divided into two groups which were not treated by smoothing and baseline correction. Spectral regions selected to build PLSR model were edited according to the qualitatively evaluating moisture sorption of heat-treated wood, which would be described in more detail in upcoming sections. Meanwhile, to avoid bias in these two groups, the original spectra collected at twelve levels of RH with five replicates (i.e., sixty spectra) were selected as calibration group. The remaining micro-FTIR spectra measured at four levels of RH with five replicates (i.e., twenty spectra) were introduced into validation group. Using these spectral data in calibration group, calculations and cross-validations were employed. After several iterations, the optical PLSR model was regressed, and the correlation of determination (R^2) and root mean square error of cross validation (RMSECV) were calculated. Then using these spectral data in validation group, the model evaluation was performed. Meanwhile, R^2 and root mean square error of prediction (RMSEP) were obtained.

2.3 Results and discussion

2.3.1 Qualitatively analyzing moisture sorption

Figure 2.4 shows heat-treated wood spectra collected at the moisture sorption process. The development of micro-FTIR spectra *vs.* RH was shown clearly.

The main band at 3358cm^{-1} assigned to O—H stretching vibration was enlarged, indicating that water was successively adsorbed on the OH group. In the spectrum collected at 0% RH, the 1739 cm^{-1} band was assigned to the C=O stretching vibration in the O=C—OH group of the glucuronic acid unit in the xylan. For comparison, the spectrum of heat-treated wood measured at 90% RH was also displayed, in which this band appeared at 1736 cm^{-1}. As the RH increased, the location of this band appeared red shift, which indicated that some water molecule was adsorbed in the C=O group. To summarize, it was confirmed that spectral regions impacted by water adsorption included 3700-3100 cm^{-1} and 1780-1700 cm^{-1}.

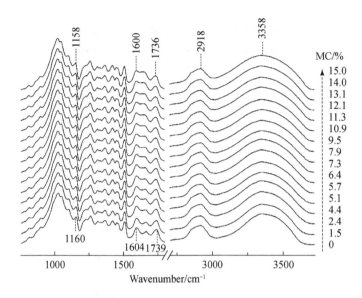

Figure 2.4　Heat-treated wood spectra collected during the moisture adsorption process

In order to further extract precise spectral information about moisture sorption, difference spectrum was employed. Figure 2.5 presents difference spectra at various RH levels acquired by subtracting the spectrum collected at 0% RH.

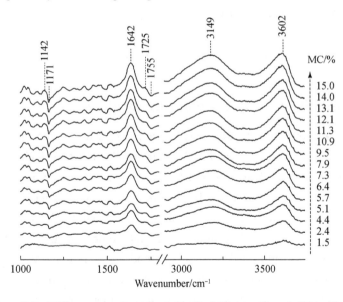

Figure 2.5　Difference spectra collected in the MC range from 1.5% to 15.0%

The broad band situated between 3700 cm^{-1} and 2987 cm^{-1} enlarged *vs.* RH which was due to moisture sorption on the OH group of wood. The same varying tendency was found in other lignocellulosic materials. Based on the above results, more accurate spectral region closely associated with moisture sorption was inferred as 3700-2987 cm^{-1}. The 1755 cm^{-1} band belonging to free hydrogen bonded C=O group decreased, while the 1725 cm^{-1} band due to one hydrogen bonded C=O group had reverse trend. Meanwhile, the band around 1642 cm^{-1} corresponded to the H—O—H bending vibration. From these, the spectral region of 1780-1525 cm^{-1} was further confirmed to be affected by water uptake. With the increase of RH, the same reflections happened in two bands located at 1171 cm^{-1} and 1142 cm^{-1}. Accordingly, the spectral region situated between 1178 cm^{-1} and 1134 cm^{-1} was also affected by water adsorption. To summarize, spectral regions closely related to water adsorption included 3700-2987 cm^{-1}, 1780-1525 cm^{-1}, and 1178-1134 cm^{-1}.

Further, the peak height changes of three bands impacted by moisture sorption with the increase of RH are displayed in Figure 2.6. It could be clearly observed that the growth trends of these three peaks *vs.* RH were different. Similar differences between these peaks closely related to moisture sorption were confirmed in other lignocellulosic materials. Therefore, any of these peaks failed to determine the moisture sorption isotherm using the univariate analysis. This might be attributable to the fact that all sorption sites played their part during the moisture sorption process, rather than only one sorption site. Considering this reason, multivariate analysis model for instance PLSR approach, should be suitable for quantitatively evaluating moisture sorption heat-treated wood.

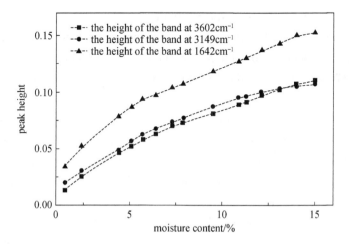

Figure 2.6 The variation in the peak height for three peaks impacted by moisture sorption *vs.* MC

2.3.2 Quantitative analysis of moisture sorption

1. Obtaining the reference value using DVS apparatus

DVS has offered vast amounts of moisture sorption isotherms. Hence, this technique was introduced to collect MC as reference values. The recorded sorption isotherm is shown in Figure 2.7, where the MC during a wide range of RH from 0 to 95% can be gained from experiment and interpolation. This isotherm curve presented the typical sigmoidal shape commonly observed for other lignocellulosic materials.

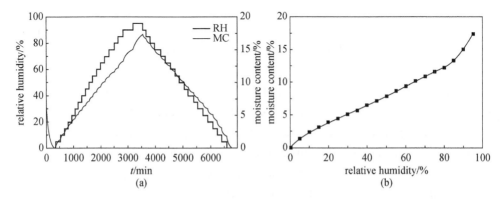

Figure 2.7 (a) Change in moisture content of heat-treated wood with the varying RH levels over the time profile in the isotherm run; (b) Equilibrium moisture content of heat-treated wood over the full set RH range in the adsorption process

2. Building the PLSR quantitative model

In order to establish PLSR model for quantitative analysis of moisture sorption, the essential argument of spectral region should be determined firstly. The spectral regions of 3700-2987cm^{-1}, 1780-1580cm^{-1}, and 1178-1134 cm^{-1} closely associated with water uptake as previously described were proposed as the first case. Based on that, the widened and narrowed spectral regions were recommended separately as the second and third case (the second case: 3700-2987 cm^{-1}, 2987-2455 cm^{-1}, 1780-1525 cm^{-1}, and 1178-1134 cm^{-1}; the third case: 3700-3100 cm^{-1}, 1780-1525 cm^{-1}, and 1178-1134 cm^{-1}). For comparison, three cases were all introduced to build the PLSR quantitative model. For these three cases, the setting parameters such as selected spectral regions and the quality parameters including RESECV, RESEP, and R^2 were all summarized in Table 2.1. The developed model in the first case owned maximum R^2, minimum RMSEP and RMSECV. It was confirmed that the first model possessed high forecast precision. The best model took advantage of the complete spectral regions related to moisture sorption (3700-2987 cm^{-1}, 1780-1525 cm^{-1}, and 1178-1134 cm^{-1}), while the widened and narrowed spectral regions (the second case and third case) influenced the performance of the established model.

Table 2.1 PLSR quality parameters for cross and test set validation for the proposed three cases

	First case	Second case	Third case
Spectral region/cm^{-1}	3700-2800/1770-1580 /1180-1140	3700-2800/2800-2700/ 1770-1580/1180-1140	3700-3000/1770-1580 /1180-1140
Cross validation			
RMSECV/%	50.5	73.6	54.3
Number of PLS components	6	6	6
R^2	0.9945	0.9898	0.9937
External validation			
RMSEP/%	19.3	28.9	25.2
Number of PLS components	6	6	6
R^2	0.9988	0.9982	0.9976

RMSECV: root mean square error of cross validation; RMSEP: root mean square error of prediction; R^2: coefficient of determination.

Taking advantage of the best model, MCs of heat-treated wood were estimated. For comparison, the reference values measured by DVS apparatus were also displayed in Figure 2.8. As described in the previous literature, the adsorption process of heat-treated wood was divided into three sections. In the first section, the absorbed water was bound directly to the hydroxyl groups and carboxyl groups, and this process followed the Langmuir's model. In the second section, the first hydration layer of these effective adsorption sites was almost fully occupied, and the absorbed water was indirectly bond via another water molecule. In the third section, absorbed water interacted with one another, and formed the five-molecule tetrahedral structure. In each section, the predicted moisture contents based on micro-FTIR spectra were much closed to the reference values (relative error was lower than 3%). The relative errors were distinguishing between three sections. This could be due to sorption mechanism. In conclusion, based on these results, it could be concluded that this proposed detection method of moisture content in nanogram quantities of heat-treated wood based on micro-FTIR spectroscopy and partial least squares regression was effective and efficient. Comparing the traditional DVS, it has unique advantage of rapid analysis and less sample consumption (ng level).

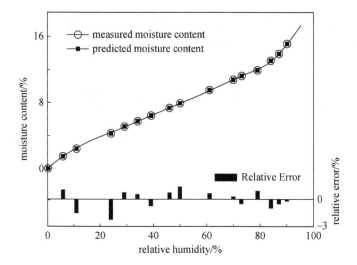

Figure 2.8 Moisture sorption isotherm estimated by micro-FTIR forecasting model and measured by DVS approach

2.4 Conclusions

One rapid and quantitative detection method of moisture content in nanogram quantities of heat-treated wood based on micro-FTIR spectroscopy and partial least squares regression was proposed here. Micro-FTIR spectra of nanogram-scaled wood were collected during the RH increasing from 0% to 90%. The analysis of these spectra and corresponding difference spectra showed that OH and C=O groups were effective water adsorption sites. It was also identified that three micro-FTIR spectral regions, such as 3700-2987 cm^{-1}, 1780-1525 cm^{-1}, and 1178-1134 cm^{-1} were closely associated with moisture sorption of heat-treated wood. Moreover, based on these identified spectral regions and MC of wood measured by DVS technique, three multivariate quantitative models predicting moisture contents were constructed using PLSR approach. By comparing these constructed PLSR models, it was summarized that the optical model was based on these three identified micro-FTIR spectral regions. Further, this optical model was used to determine moisture sorption isotherm of wood which matched that measured using DVS apparatus, and it confirmed the validity in the detection method of moisture content in nanogram quantities of heat-treated wood based on micro-FTIR spectroscopy and partial least squares regression.

References

AKGÜL M, GÜMÜŞKAYA E, KORKUT S. 2007. Crystalline structure of heat-treated Scots pine [Pinus sylvestris L.] and Uludağ fir [Abies nordmanniana (Stev.) subsp. Bornmuelleriana (Mattf.)] wood [J]. Wood Science and Technology, 41 (3): 281-289.

ALMEIDA G, BRITO J O, PERRE P. 2009. Changes in wood-water relationship due to heat treatment assessed on micro-samples of three Eucalyptus species [J]. Holzforschung, 63 (1): 80-88.

ATALLA R H. 1987. Raman spectroscopy and the Raman microprobe: valuable new tools for characterizing wood and wood pulp fibers [J]. Journal of Wood Chemistry and Technology, 7 (1): 115-131.

AWOYEMI L, COOPER P A, UNG T Y. 2009. In-treatment cooling during thermal modification of wood in soy oil medium: soy oil uptake, wettability, water uptake and swelling properties [J]. European Journal of Wood and Wood Products, 67 (4): 465-470.

BARDET M, FORAY M F, MARON S, et al. 2004. Characterization of wood components of Portuguese medieval dugout canoes with high-resolution solid-state NMR [J]. Carbohydrate Polymers, 57 (4): 419-424.

BELFER S, FAINCHTAIN R, PURINSON Y, et al. 2000. Surface characterization by FTIR-ATR spectroscopy of polyethersulfone membranes-unmodified, modified and protein fouled [J]. Journal of Membrane Science, 172 (1-2): 113-124.

BENAR P, GONÇALVES A R, MANDELLI D, et al. 1999. Principal component analysis of the hydroxymethylation of sugarcane lignin: a time-depending study by FTIR [J]. Journal of Wood Chemistry and Technology, 19 (1-2): 151-165.

BESKERS T F, HOFE T, WILHELM M. 2015. Development of a chemically sensitive online SEC detector based on FTIR spectroscopy [J]. Polymer Chemistry, 6 (1): 128-142.

BOONSTRA M J, TJEERDSMA B. 2006. Chemical analysis of heat treated softwoods [J]. Holz als Roh und Werkstoff, 64 (3): 204-211.

BORREGA M, KARENLAMPI P P. 2010. Hygroscopicity of heat-treated Norway spruce (Picea abies) wood [J]. European Journal of Wood and Wood Products, 68 (2): 233-235.

CASIERI C, SENNI L, ROMAGNOLI M, et al. 2004. Determination of moisture fraction in wood by mobile NMR device [J]. Journal of Magnetic Resonance, 171 (2): 364-372.

COLOM X, CARRILLO F. 2005. Comparative study of wood samples of the northern area of catalonia by FTIR [J]. Journal of Wood Chemistry and Technology, 25 (1-2): 1-11.

DILIK T, HIZIROGLU S. 2012. Bonding strength of heat treated compressed Eastern redcedar wood [J]. Materials and Design, 42: 317-320.

DING T, WANG C, PENG W. 2016. A theoretical study of moisture sorption behavior of heat-treated pine wood using Raman spectroscopic analysis [J]. Journal of Forestry Engineering, 1 (5): 15-19.

ENGELUND E T, THYGESEN L G, SVENSSON S, et al. 2013. A critical discussion of the physics of wood-water interactions [J]. Wood Science and Technology, 47 (1): 141-161.

ESTEBAN L G, FERNANDEZ F G, CASASUS A G, et al. 2006. Comparison of the hygroscopic behavior of 205-year-old and recently cut juvenile wood from Pinus sylvestris L. [J]. Annals of Forest Science, 63 (3): 309-317.

ESTEVES B, PEREIRA H. 2008. Quality assessment of heat-treated wood by NIR spectroscopy [J]. Holz Als Roh-und Werkstoff, 66 (5): 323-332.

EVANS P D, BANKS W B. 1988. Degradation of wood surfaces by water changes in mechanical-properties of thin wood strips [J]. Holz Als Roh-und Werkstoff, 46 (11): 427-435.

FERRAZ A, MENDONCA R, GUERRA A, et al. 2005. Near-infrared spectra and chemical characteristics of pinus taeda (Loblolly pine) wood chips biotreated by the white rot fungus ceriporiopsis subvermispora [J]. Journal of Wood Chemistry and Technology, 24 (2): 99-113.

GUO X, WU Y. 2018. *In situ* visualization of water adsorption in cellulose nanofiber film with micrometer spatial resolution using micro-FTIR imaging [J]. Journal of Wood Chemistry and Technology, 38 (5): 361-370.

GUO X, WU Y, YAN N. 2017. Characterizing spatial distribution of the adsorbed water in wood cell wall of Ginkgo biloba L. by mu-FTIR and confocal Raman spectroscopy [J]. Holzforschung, 71 (5): 415-423.

HILL C A S, NORTON A, NEWMAN G. 2010. The water vapor sorption behavior of flax fibers—analysis using the parallel exponential kinetics model and determination of the activation energies of sorption [J]. Journal of Applied Polymer Science, 116 (4): 2166-2173.

HILL C A S, NORTON A, NEWMAN G. 2010. Analysis of the water vapor sorption behavior of Sitka spruce [Picea sitchensis (Bongard) Carr.] based on the parallel exponential kinetics model [J]. Holzforschung, 64 (4): 469-473.

HILL C A S, RAMSAY J, KEATING B, et al. 2012. The water vapor sorption properties of thermally modified and densified wood [J]. Journal of Materials Science, 47 (7): 3191-3197.

HOGAN JR C J, NIKLAS K J. 2004. Temperature and water content effects on the viscoelastic behavior of Tilia americana(Tiliaceae) sapwood [J]. Trees, 18 (3): 339-345.

HSI E, HOSSFELD R, BRYANT R G. 1977. Nuclear magnetic resonance relaxation study of water absorbed on milled northern white-cedar [J]. Journal of Colloid and Interface Science, 62 (3): 389-395.

HUANG X, KOCAEFE D, KOCAEFE Y, et al. 2012. Changes in wettability of heat-treated wood due to artificial weathering [J]. Wood Science and Technology, 46 (6): 1215-1237.

INAGAKI T, YONENOBU H, TSUCHIKAWA S. 2008. Near-infrared spectroscopic monitoring of the water adsorption/desorption process in modern and archaeological wood [J]. Applied Spectroscopy, 62 (8): 860-865.

IRVINE J, GRACE J. 1997. Continuous measurements of water tensions in the xylem of trees based on the elastic properties of wood [J]. Planta, 202 (4): 455-461.

ITO Y, NAKASHIMA S. 2002. Water distribution in low-grade siliceous metamorphic rocks by micro-FTIR and its relation to grain size: a case from the Kanto Mountain region, Japan [J]. Chemical Geology, 189 (1-2): 1-18.

JACKSON P, DENT G, CARTER D et al. 1993. Investigation of high sensitivity GC‐FTIR as an analytical tool for structural identification [J]. Journal of High Resolution Chromatography, 16 (9): 515-521.

JALALUDIN Z, HILL C A S, XIE Y, et al. 2010. Analysis of the water vapor sorption isotherms of thermally modified acacia and sesendok [J]. Wood Material Science and Engineering, 5 (3-4): 194-203.

JIANG X, LI S, XIANG G, et al. 2016. Determination of the acid values of edible oils via FTIR spectroscopy based on the O—H stretching band [J]. Food Chemistry, 212: 585-589.

JOMA E, SCHMIDT G, CREMONEZ V G, et al. 2017. The effect of heat treatment on wood-water relationship and mechanical properties of commercial uruguayan plantation timber eucalyptus grandis [J]. Australian Journal of Basic and Applied Sciences, 10 (1): 704-708.

KARTAL S N, HWANG W, IMAMURA Y. 2007. Water absorption of boron-treated and heat-modified wood [J]. Journal of Wood Science, 53 (5): 454-457.

KEKKONEN P M, YLISASSI A, TELKKI V. 2014. Absorption of water in thermally modified pine wood as studied by nuclear magnetic resonance [J]. Journal of Physical Chemistry C, 118 (4): 2146-2153.

KOTILAINEN R A, TOIVANEN T J, ALEN R J. 2000. FTIR monitoring of chemical changes in softwood during heating [J]. Journal of Wood Chemistry and Technology, 20 (3): 307-320.

LESTANDER T A. 2008. Water absorption thermodynamics in single wood pellets modelled by multivariate near-infrared spectroscopy [J]. Holzforschung, 62 (4): 429-434.

LING S, QI Z, KNIGHT D P, et al. 2013. FTIR imaging, a useful method for studying the compatibility of silk fibroin-based polymer blends [J]. Polymer Chemistry, 4 (21): 5401-5406.

MAEDA H, FUKADA E. 1987. Effect of bound water on piezoelectric, dielectric, and elastic properties of wood [J]. Journal of Applied Polymer Science, 33 (4): 1187-1198.

METSÄ-KORTELAINEN S, ANTIKAINEN T, VIITANIEMI P. 2006. The water absorption of sapwood and heartwood of Scots pine and Norway spruce heat-treated at 170 ℃, 190 ℃, 210 ℃ and 230 ℃ [J]. Holz als Roh- und Werkstoff, 64 (3): 192-197.

MITSUI K, INAGAKI T, TSUCHIKAWA S. 2008. Monitoring of hydroxyl groups in wood during heat treatment using NIR spectroscopy [J]. Biomacromolecules, 9 (1): 286-288.

MOLIŃSKI W, RACZKOWSKI J. 1988. Mechanical stresses generated by water adsorption in wood and their determination by tension creep measurements [J]. Wood Science and Technology, 22 (3):

193-198.

OH S Y, YOO D I, SHIN Y, et al. 2005. FTIR analysis of cellulose treated with sodium hydroxide and carbon dioxide [J]. Carbohydrate Research, 340 (3): 417-428.

OLEK W, MAJKA J, CZAJKOWSKI L. 2013. Sorption isotherms of thermally modified wood [J]. Holzforschung, 67 (2): 183-191.

OUERTANI S, AZZOUZ S, HASSINI L, et al. 2014. Moisture sorption isotherms and thermodynamic properties of Jack pine and palm wood: Comparative study [J]. Industrial Crops and Products, 56: 200-210.

OZGENC O, DURMAZ S, BOYACI I H, et al. 2017. Determination of chemical changes in heat-treated wood using ATR-FTIR and FT Raman spectrometry [J]. Spectrochimica Acta Part A-Molecular and Biomolecular Spectroscopy, 171: 395-400.

PANDEY S, PANDEY P, TIWARI G, et al. 2012. FTIR spectroscopy: a tool for quantitative analysis of ciprofloxacin in tablets [J]. Indian Journal of Pharmaceutical Sciences, 74 (1): 86-90.

PENTRAK M, BIZOVSKA V, MADEJOVA J. 2012. Near-IR study of water adsorption on acid-treated montmorillonite [J]. Vibrational Spectroscopy, 63: 360-366.

POPESCU C M, HILL C A S, POPESCU M C. 2016. Water adsorption in acetylated birch wood evaluated through near infrared spectroscopy [J]. International Wood Products Journal, 7 (2): 61-65.

SALCA E, HIZIROGLU S. 2014. Evaluation of hardness and surface quality of different wood species as function of heat treatment [J]. Materials and Design, 62: 416-423.

SAMUEL E, MOHAN S. 2004. FTIR and FT Raman spectra and analysis of poly(4-methyl-1-pentene) [J]. Spectrochimica Acta Part A-Molecular and Biomolecular Spectroscopy, 60 (1-2): 19-24.

SANDAK A, SANDAK J, ALLEGRETTI O. 2015. Quality control of vacuum thermally modified wood with near infrared spectroscopy [J]. Vacuum, 114: 44-48.

SCHEIDING W, DIRESKE M, ZAUER M. 2016. Water absorption of untreated and thermally modified sapwood and heartwood of Pinus sylvestris L. [J]. European Journal of Wood and Wood Products, 74 (4): 585-589.

SEARSON M J, THOMAS D S, MONTAGU K D, et al. 2004. Wood density and anatomy of water-limited eucalypts [J]. Tree Physiology, 24 (11): 1295-1302.

SENNI L, CASIERI C, BOVINO A, et al. 2009. A portable NMR sensor for moisture monitoring of wooden works of art, particularly of paintings on wood [J]. Wood Science and Technology, 43 (1-2): 167-180.

SRINIVAS K, PANDEY K K. 2012. Effect of heat treatment on color changes, dimensional stability, and mechanical properties of wood [J]. Journal of Wood Chemistry and Technology, 32 (4): 304-316.

SUN B, WANG Z, LIU J. 2017. Changes of chemical properties and the water vapor sorption of

Eucalyptus pellita wood thermally modified in vacuum [J]. Journal of Wood Science, 63 (2): 133-139.

TUROV V V, LEBODA R. 1999. Application of ^{1}H NMR spectroscopy method for determination of characteristics of thin layers of water adsorbed on the surface of dispersed and porous adsorbents. [J]. Advances in Colloid and Interface Science, 79 (2): 173-211.

WANG S, MAHLBERG R, JAMSA S, et al. 2011. Surface properties and moisture behavior of pine and heat-treated spruce modified with alkoxysilanes by sol-gel process [J]. Progress in Organic Coatings, 71 (3): 274-282.

WIBERG P, MORÉN T J. 1999. Moisture flux determination in wood during drying above fibre saturation point using CT-scanning and digital image processing [J]. Holz Als Roh-und Werkstoff, 57 (2): 137-144.

WILLEMS W. 2014. The water vapor sorption mechanism and its hysteresis in wood: The water/void mixture postulate [J]. Wood Science and Technology, 48 (3): 499-518.

WILLEMS W, ALTGEN M, MILITZ H. 2015. Comparison of EMC and durability of heat treated wood from high versus low water vapor pressure reactor systems [J]. International Wood Products Journal, 6 (1): 21-26.

ZELINKA S L, GLASS S V, STONE D S. 2008. A percolation model for electrical conduction in wood with implications for wood water relations [J]. Wood and Fiber Science, 40 (4): 544-552.

ZHANG M H. 2011. Mechanism of water sorption during adsorption process of wood studied by NMR [J]. Chinese Journal of Magnetic Resonance, 28 (1): 135-141.

ZHANG X, KUNZEL H M, ZILLIG W, et al. 2016. A Fickian model for temperature-dependent sorption hysteresis in hygrothermal modeling of wood materials [J]. International Journal of Heat and Mass Transfer, 100: 58-64.

Chapter 3 Quantitative analysis of moisture sorption in lignin using micro-FTIR spectroscopy

3.1 Introduction

With the increasing shortage of fossil resources, the shortage of chemicals based on fossil resources is also coming, and the use of plant biomass-based materials to supplement petroleum-based polymer materials has become increasingly important. Lignin, a typical plant biomass-based material, like cellulose and hemicellulose, is a natural polymer present in plant cell walls (Ge et al. 2016; Hu and Hsieh 2016; Kilpeläinen et al. 1994; Thakur and Thakur 2015). This polymer is collected from the pulp and paper industry (Aghdam et al. 2016; De Los Santos Ramos et al. 2009; Song et al. 2016) and because of its high carbon content (typically between 55% and 66%), it can be used as a carbon fiber raw material. At the same time, lignin contains various functional groups such as hydroxyl and ether bonds, which can generate strong intermolecular forces with polymers to prepare composite materials by blending with resins. In addition, lignin has been widely used as biopolymer additive in material manufacturing, including for flame retardant (Costes et al. 2016; De Chirico et al. 2003; Fu and Cheng 2011; Song et al. 2011; Yu et al. 2012) and adhesive additives (Aracri et al. 2014; Khan et al. 2004; Koumba-Yoya and Stevanovic 2017; Rahman et al. 2016), as well as dispersing (Milczarek 2010; Stewart 2008), plasticization (Bouajila et al. 2005; Feldman et al. 2007) and coupling agents (Fernandes et al. 2014; Rozman et al. 2001; Xu et al. 2015). In recent years, it has been integrated into polymeric matrixes or nanoparticles to form lignin-based material with novel functions (e.g., optical absorption in the ultraviolet spectrum) (Belov et al. 2011; Xiong et al. 2018), high mechanical strength (Bouajila et al. 2005; Nair et al. 2017; Nitz et al. 2001; Warth et al. 1997; Yoshida et al. 1987), and oxidation resistance (Faustino et al. 2010; Kai et al. 2016; Kasprzycka-Guttman and Odzeniak 1994; Pouteau et al. 2003; Sadeghifar and Argyropoulos 2015). The commercial use of lignin to produce lignin-based material with certain desirable physical and chemical properties remains challenging, as many of these properties should be further studied. Among them, hygroscopic behavior is a key property. Lignin is inherently hygroscopic because it contains many hydrophilic groups (Muraille et al. 2015; Pepper et al. 1959; Rawat and Khali 1999; Shaha et al. 2011), such as hydroxyl (phenolic or alcoholic), methoxyl, carbonyl, and carboxyl groups, which may affect the performance of lignin and lignin-based material (Boeriu et al. 2004; Celino et al. 2014; Vu et al. 2002). Taking this into consideration, understanding moisture sorption mechanisms of lignin is extremely essential.

Moisture sorption, an important property of lignin and lignin-based materials, has been studied from various perspectives, including the sorption isotherm (Passauer et al. 2012; Reina et al. 2001; Volkova et al. 2012). Rawat *et al.* (Rawat and Khali 1999) determined the sorption isotherm of lignin and evaluated the isotherm data via the Brunauer-Emmett-Teller (BET) model. Furthermore, Volkova et al. (Volkova et al. 2012) studied sorption behavior of Kraft lignin using a dynamic vapor sorption (DVS) apparatus and analyzed the sorption data using BET and Guggenheim-Anderson-de Boer models. Meanwhile, Reina et al. (Reina et al. 2001) studied the role of lignin in the moisture sorption of conifer cuticles and showed that the lignin fraction had both high water sorption and the capability of retaining water in comparison to the cuticle component. Stiubianu *et al.* (Stiubianu et al. 2011) examined moisture sorption of lignin and polydimethylsiloxane composites and found that the incorporation of lignin changed the water sorption capacity of lignin-based material due to the presence of the lignin polar hydrophilic groups. In addition, the sorption isotherms and extent of sorption hysteresis of other lignin-based materials were determined using the DVS method, including for lignin-based activated carbons, lignin-cellulose hydrogels, poly (ethylene) glycol diglycidyl ether-modified lignin xerogel, and thermally modified wood. Although the DVS method can provide moisture sorption data for lignin and lignin-based materials, it still suffers from resolution-related issues regarding the size (centimeters) as well as mass of the measured samples (milligrams) (Passauer et al. 2012; Spiridon et al. 2015). Therefore, developing a new approach with higher resolution is urgently required.

Meanwhile, the molecular interaction between lignin and water is one of the most important factors affecting moisture adsorption within lignin and lignin-based materials. As mentioned, lignin has a highly complex structure containing several potential water sorption sites (Dababi et al. 2016). Due to various hydrogen bond formations based on different sorption sites and the molecular structure of adsorbed water differing from general water, the moisture sorption mechanism of lignin is highly complex and requires further qualitative analysis. Fourier transform infrared (FTIR) spectroscopy has been adapted to study this phenomenon since it allows characterization of molecular structure (Popescu et al. 2010). For example, Olsson *et al.* (Olsson and Salmen 2004) observed the moisture adsorption characteristics of wood polymers by FTIR spectroscopy at nine levels of relative humidity (RH) ranging from 0% to 80%, subsequently described the water sorption sites, and determined an important relationship between weight gain and increase of the O — H stretching band. Balcerzak *et al.* (Balcerzak and Mucha 2008) confirmed that the absorbance of the hydroxyl group band was increased with an increase in adsorbed water. Furthermore, Célino *et al.* (Celino et al. 2014) and Xu *et al.* (Xu et al. 2013) showed that the spectral information for FTIR spectra could be used to qualitatively and quantitatively analyze the moisture absorption mechanisms of lignin fibers. Indeed, this technique provides some obvious advantages, i.e., the easily distinguishable characteristic peak (Boeriu et al. 2004; Faix 1991), the high sensitivity of water (Meng et al. 2012), and high accuracy of spectral analysis (Moreira and Santos 2004). More recently, micro-FTIR spectroscopy equipped with an additional visible-light microscope offering visual examination and selecting observation area has been developed. This improved technique yields high spatial resolution at the micron scale (Liu et al. 2008) and can provide *in situ* observation of water adsorption (Guo et al. 2018), which may feature multi-parameter measurements, rapid analysis, lower sample consumption, and high resolution (Dean et al. 2010; Guo et al. 2017).

As previously described, micro-FTIR spectroscopy is an effective technique for observing moisture adsorption. Therefore, in this study, an *in situ* approach was developed for qualitative and quantitative analysis of moisture adsorption in ng-scaled lignin by using micro-FTIR spectroscopy and partial least squares regression. First, we measured micro-FTIR spectra of ng-scaled lignin over a wide range of RH levels, which could be accurately controlled by a specially designed apparatus. Second, we qualitatively analyzed these spectra along with the corresponding difference spectra to confirm effective water sorption sites and identify the spectral ranges related to moisture adsorption. Third, using the partial least squares regression, we built and developed multivariate forecasting models based on the identified spectral ranges and measured moisture contents (MCs) using the DVS technique. Finally, through comparative analyses between the estimated and measured MCs, we illustrated the practicability and effectiveness of the proposed *in situ* qualitative and quantitative analysis approach.

3.2 Materials and methods

3.2.1 Materials

The commercial lignin was obtained from Sigma-Aldrich Co., Ltd. as a raw material brown powder. The C=O content within the lignin was determined according to the literature (El Mansouri and Salvado 2007; Erdocia et al. 2014; Faix et al. 1998; Yanez-S et al. 2014; Zhu et al. 2014). The measurement was repeated three times, and the average of the C=O content was 3.22%. Then the 1 wt. % suspension of lignin was prepared from this raw material. Droplets of this suspension were sprayed onto the bottom of the sample cell from an injector. The droplet size ranged from 5 to 20 microns (~0.5-33 ng). 10h later, the water in the droplets had evaporated, and only lignin film was left (~1 ng). The sample cell was then sealed for spectral measurement.

3.2.2 Experimental apparatus for micro-FTIR spectral measurement

The experimental apparatus is presented in Figure 3.1(a). The primary component was a spectrometer (Nicolet IN 10^{TM}), which was applied for recording micro-FTIR spectral development of lignin vs. RH. This spectrometer was installed with an additional microscope for visual examination. During the spectral measurement, standard spectral acquisition (spectral collection at the selected point) was adopted, and all micro-FTIR spectra were acquired from one randomly selected area (50 μm×50 μm), in which only one lignin film was present (~1 ng). Meanwhile, these spectra in the range of 720-4000 cm^{-1} were collected 32 times by scanning the grating at the spectral resolution of 4 cm^{-1}. Figure 3.1(b) shows the diagram of sample cell. First, the lignin sample was placed on the bottom of this sample cell. Then, this cell was closed and subsequently added to the automatic stage of the micro-FTIR spectrometer. The RH was changed using a saturated steam-air mixture accurately controlled by a flow meter. At that point, the RH could be tested using a humidity and temperature meter and real-time displayed on the computer.

Throughout the experiment, all spectral measurements were carried out within the RH range from 0% to 92% at a constant temperature of 25 ℃. Representative variations of target RH, actual RH, and recorded micro-FTIR spectra of lignin over time are shown in Figure 3.2. Once the target RH was set to a new value (e.g., 5% and 10% RH), a delay of approximately 5 min would occur. During this delay, actual RH approached target RH and then remained nearly unchanged afterward. Meanwhile, after 15 min, the spectra recorded every one minute was kept constant (the peak height of the 3376 cm^{-1} band affected by moisture adsorption was used to demonstrate spectral change). Based on the above results, 60 min were set to balance the time for micro-FTIR spectral measurement.

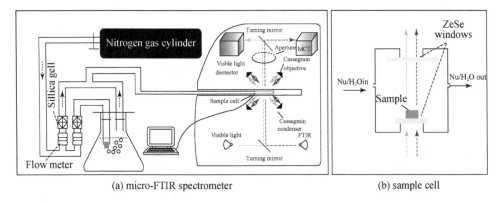

Figure 3.1 (a) Diagrammatic sketch of experimental installation; (b) Diagrammatic sketch of sample cell

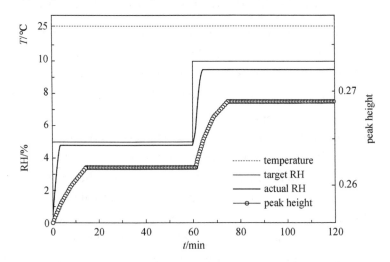

Figure 3.2 Representative variations of target RH, actual RH and micro-FTIR spectra of lignin against time

3.2.3 Experimental apparatus for moisture content measurement

A DVS apparatus (DVS AdvantagePlus) was used to measure MC. First, a lignin specimen was placed on a metal plate; then the plate was connected to a microbalance in a temperature and humidity automatically controlled box. The target RH was set to increase at steps of 5% RH up to a maximum of 95% RH, before reduction to 0% RH in the same step. During this process, the target RH would be maintained for a sufficient amount of time for the value of the weight change per minute to fall below 0.002%, and then the target RH became the new preset value. Meanwhile, the temperature was maintained at 25 °C.

The MC was computed by using the following formula:

$$\mathrm{MC} = \frac{m - m_d}{m_d} \times 100\% \qquad (3.1)$$

Where, MC was moisture content, m_d was the quality of dry sample, and m was the quality of the sample at various RH levels.

Using MC measurements obtained with the DVS apparatus, representative variations of RH and MC against time are demonstrated in Figure 3.3. Once the target RH became a new value (e.g., 5% and 10% RH), a time delay of 4-10 min would occur. Within this duration, actual RH approached target RH, then remained stable throughout the experiment. Meanwhile, typical variation of MC with time followed an asymptotic curve, and the MC achieved equilibrium value after infinite time of exposure at the target RH. More importantly, three replicates of these lignin specimens were adopted, and the average MC recorded for these three replicates was provided as a reference value.

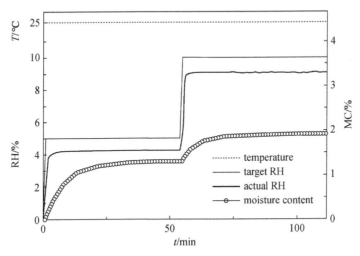

Figure 3.3 Typical variations of target RH, actual RH and MC of lignin against time

3.2.4 Micro-FTIR spectral data processing

1. Achievement of difference spectra at various RH levels

To further qualitatively analyze the moisture adsorption of lignin, an FTIR difference spectrum technique was applied, by which difference spectra were obtained via the original micro-FTIR spectra collected at different RH levels, while subtracting the spectrum measured at 0% RH.

2. Establishment of the micro-FTIR forecasting model

A multivariate quantitative forecasting model was generated using a partial least square regression (PLSR) approach in TQ Analyst 9. The path length type could be set to "constant". The primary component MC was filled first in the component table, and maximum and minimum values of MC were individually set to "16%" and "0%", respectively, which was obtained from reference value. Then the original micro-FTIR spectra of lignin collected at sixteen RH levels were divided into two groups, which were not all treated by using smoothing and baseline correction approach. Spectral ranges selected to establish a quantitative forecasting model were edited on the basis of the qualitative characterization of lignin moisture adsorption, which will be explained in the upcoming section 3.3.1. Meanwhile, to avoid bias within these two groups, the original micro-FTIR spectra collected at twelve RH levels with five replicates (i.e., sixty spectra) were selected as the calibration group. The remaining micro-FTIR spectra measured at four RH levels with five replicates (i.e., twenty spectra) were introduced into the validation group. Using the spectral data of the calibration group, calculations and cross-validations were applied. After several iterations, the optical quantitative forecasting model was obtained, and the correlation of determination (R^2) and root mean square error of cross validation (RMSECV) were calculated. Then the model evaluation was performed using spectral data within the validation group. Meanwhile, the R^2 and root mean square error of prediction (RMSEP) were obtained.

3.3 Results and discussion

3.3.1 Quantitative analysis of moisture adsorption in lignin

Figure 3.4 demonstrates the micro-FTIR spectra of lignin in the RH region of 0%-92%. In this figure, the development of micro-FTIR spectra *vs.* RH is clearly shown. As mentioned before, lignin contains functional groups, including hydroxyl (phenolic or alcoholic), methoxyl, and carbonyl groups, which are important structural characteristics. The assignment of primary bands in these recorded spectra is listed in Table 3.1.

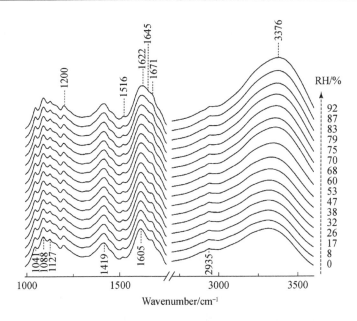

Figure 3.4 Micro-FTIR spectra of lignin collected over a wide range of RH from 0% to 92%

Table 3.1 Assignment of the main bands in micro-FTIR spectra of lignin closely associated with water adsorption

Wavenumber/cm^{-1}	Assignment
3376	O—H stretching vibration
2935	C—H stretch in methyl and methylene groups
1671	C=O stretching vibration
1645	H—O—H bending vibration
1605	aromatic skeletal plus C=O stretching vibration
1516	aromatic skeleton vibration
1419	aromatic skeletal vibration combined with C—H in-plane deformation
1200	C—C plus C—O plus C=O stretching vibration
1127	aromatic C—H in-plane deformation
1088	C—O deformation
1041	aromatic C—H in-plane deformation plus C—O deformation plus C=O stretch

With an increase in RH, one key band near 3376 cm^{-1} corresponding to the O—H stretching vibration was enlarged, indicating that water molecule was successively adsorbed in the OH group. In the spectrum collected at 92% RH, the 1671 cm^{-1} and 1200 cm^{-1} bands belonged to C=O stretching vibration and C—C plus C—O plus C=O stretching vibration, respectively. These two bands arise due to lignin macromolecule moisture adsorption and suggested that the C=O group was also the water adsorption site of lignin. The 1622 cm^{-1} band belonged to an aromatic skeletal plus C=O stretching vibration. For comparison, the spectrum of lignin measured at 0% RH was also evaluated, in which this band appeared at 1605 cm^{-1}. When the RH increased, this band position exhibited a blue shift, which indicated that some water molecule was adsorbed on the C=O group. Meanwhile, the bands at 2935 cm^{-1}, 1516 cm^{-1}, 1419 cm^{-1}, 1127 cm^{-1}, 1088 cm^{-1} and 1041 cm^{-1} belonged to C—H stretch in methyl and methylene groups, aromatic skeleton vibration, aromatic skeletal vibration combined with C—H in-plane deformation, aromatic C—H in-plane deformation, C—O deformation, and aromatic C—H in-plane deformation plus C—O deform plus C=O stretch. Consequently, the spectral ranges of 3700-3000 cm^{-1}, 1740-1488 cm^{-1}, and 1247-1175 cm^{-1} were closely correlated with moisture adsorption.

To further extract precise spectral information concerning moisture adsorption, a difference spectrum was employed. Figure 3.5 presents difference spectra acquired by subtracting micro-FTIR spectrum collected at 0% RH.

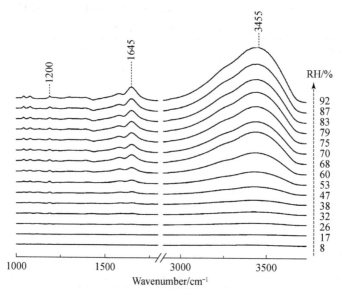

Figure 3.5　Difference micro-FTIR spectra of lignin collected during the RH increased from 8% to 92%

The broad band situated near 3455 cm^{-1} was enlarged vs. RH, which was due to water adsorption within the OH group of lignin. The same change tendency of this band was also found in other lignin material. Based on this observation, it was inferred that a more accurate spectral range of this primary band was correlated with moisture adsorption from 3700 cm^{-1} to 2880 cm^{-1}. Meanwhile, the intense band around 1645 cm^{-1} increased in all spectra corresponding to the H — O — H bending vibration. Meanwhile, the 1200 cm^{-1} band correlated with moisture adsorption displayed the same trend.

Furthermore, the peak height changes of the three bands related to moisture adsorption with an increase of RH are displayed in Figure 3.6. It can be clearly observed that the growth trends of these three bands against RH are different. Thus, none of these bands allowed determination of the water sorption isotherm using univariate analysis. This might be due to the fact that all sorption sites played their part in moisture sorption, rather than only at one sorption site. For this reason, a multivariate analysis model should be suitable for quantitatively evaluating moisture adsorption in lignin.

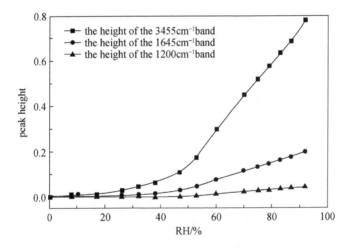

Figure 3.6 Peak height changes of micro-FTIR bands affected by moisture adsorption against RH

3.3.2 Quantitative evaluation of moisture adsorption in lignin

1. Obtaining the reference value using a DVS apparatus

As previously shown, DVS is able to offer MC as reference value for lignin moisture adsorption. The recorded sorption isotherm is illustrated in Figure 3.7, where the MC can be clearly drawn. As expected, this MC curve against RH exhibited a typical sigmoidal shape.

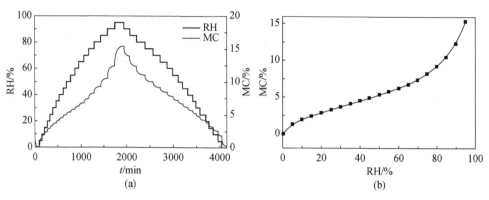

Figure 3.7 (a) Changes in MCs of lignin with the variable over the time profile during isotherm runs; (b) MCs of lignin measured using DVS technique

2. Building a micro-FTIR forecasting model

For building the forecasting model, the essential argument of spectral range should be first established. The micro-FTIR spectral ranges of 3700-2880 cm^{-1}, 1740-1488 cm^{-1}, and 1247-1175 cm^{-1} closely correlated with moisture adsorption were initially proposed. Meanwhile, either widened or narrowed spectral ranges were recommended as second and third options or cases, respectively (second case: 3700-2880 cm^{-1}, 2880-2647 cm^{-1}, 1740-1488 cm^{-1}, and 1247-1175 cm^{-1}; third case: 3700-3000 cm^{-1}, 1740-1488 cm^{-1}, and 1247-1175 cm^{-1}). For comparison, three cases were all introduced to build the micro-FTIR forecasting model. For these three cases, the setting parameters such as selected spectral ranges and the quality parameters, including RMSECV, RMSEP, and R^2 are summarized in Table 3.2. The developed forecasting model in the first case offered maximum R^2, minimum RMSEP and RMSECV. It was confirmed that the first model possessed a high forecast precision. The best model took advantage of the complete spectral ranges related to moisture adsorption (3700-2880 cm^{-1}, 1740-1488 cm^{-1}, and 1247-1175 cm^{-1}), whereas widened and narrowed spectral ranges (second and third cases) had a harmful effect on the performance of the established forecasting model.

Table 3.2 PLSR quality parameters for cross and test set validation for the proposed three cases

	First case	Second case	Third case
Spectral range/cm^{-1}	3700-2880/1740-1488/ 1247-1175	3700-2880/2880-2647/ 1740-1488/1247-1175	3700-3000/1740-1488/ 1247-1175
Cross validation			
RMSECV	0.408	0.617	0.447
Number of PLS components	4	4	4
R^2/%	99.36	98.43	99.17

Continued

	First case	Second case	Third case
External validation			
RMSEP	0.118	0.171	0.146
Number of PLS components	4	4	4
R^2/%	99.96	99.89	99.90

RMSECV: root mean square error of cross validation; R^2: coefficient of determination; RMSEP: root mean square error of prediction.

Taking advantage of the quantitative forecasting model, lignin MCs at various RH levels were estimated. For comparison, the reference values measured by DVS apparatus are also shown in Figure 3.8. The moisture adsorption of lignin-based material included three parts. In the first part, absorbed water was bound directly to the hydroxyl and carboxyl groups, with this process following the Langmuir's model. In the second part, the first hydration layer of these effective water adsorption sites was almost fully occupied, and absorbed water was indirectly bound via another water molecule. In the third part, absorbed water owned the five-molecule tetrahedral structure interacted themselves. In these three sections, the estimated MCs based on micro-FTIR spectra were quite close to the measured MCs using the DVS setup (relative error < 3%). Results indicated that this approach for *in situ* qualitative and quantitative analysis of moisture adsorption in nanogram-scaled lignin via micro-FTIR spectroscopy and partial least squares regression was effective and efficient.

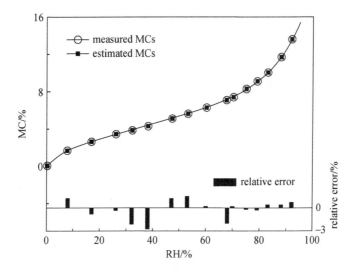

Figure 3.8 A comparison of sorption isotherm estimated by micro-FTIR forecasting model and determined by DVS technique

3.4 Conclusions

Using micro-FTIR spectroscopy and partial least squares regression, the approach for *in situ* qualitative and quantitative analysis of moisture adsorption in nanogram-scaled lignin was demonstrated. A qualitative analysis of these spectra of nanogram-scaled lignin collected in the RH region of 0%-92% and difference spectra showed that the OH and C=O groups were effective water sorption sites. Meanwhile, it was confirmed that three micro-FTIR spectral ranges, such as 3700-2880 cm^{-1}, 1740-1488 cm^{-1}, and 1247-1175 cm^{-1} are closely correlated with moisture adsorption. Moreover, based on these identified spectral ranges and the lignin MC measured by the DVS setup, the micro-FTIR forecasting model was constructed. Furthermore, this micro-FTIR forecasting model was used to generate a sorption isotherm of lignin, which matched appropriately that measured using the DVS apparatus, and confirmed the practicability and effectiveness of this *in situ* qualitative and quantitative analysis approach.

References

AGHDAM M A, KARIMINIA H, SAFARI S. 2016. Removal of lignin, COD, and color from pulp and paper wastewater using electrocoagulation [J]. Desalination and Water Treatment, 57 (21): 9698-9704.

ARACRI E, DIAZ BLANCO C, TZANOV T. 2014. An enzymatic approach to develop a lignin-based adhesive for wool floor coverings [J]. Green Chemistry, 16 (5): 2597-2603.

BALCERZAK J, MUCHA M. 2008. Study of adsorption and desorption heats of water in chitosan and its blends with hydroxypropylcellulose [J]. Molecular Crystals and Liquid Crystals, 484 (1): 465-472.

BELOV N P, SHERSTOBITOVA A S, YASKOV A D. 2011. Diffuse reflection of light by cellulose pulp and optical absorption of aqueous residual lignin solutions [J]. Journal of Applied Spectroscopy, 78 (1): 138-140.

BOERIU C G, BRAVO D, GOSSELINK R, et al. 2004. Characterisation of structure-dependent functional properties of lignin with infrared spectroscopy [J]. Industrial Crops and Products, 20 (2): 205-218.

BOUAJILA J, LIMARE A, JOLY C, et al. 2005. Lignin plasticization to improve binderless fiberboard mechanical properties [J]. Polymer Engineering and Science, 45 (6): 809-816.

CÉLINO A, GONCALVES O, JACQUEMIN F, et al. 2014. Qualitative and quantitative assessment of water sorption in natural fibers using ATR-FT'IR spectroscopy [J]. Carbohydrate Polymers, 101:

163-170.

COSTES L, LAOUTID F, AGUEDO M, et al. 2016. Phosphorus and nitrogen derivatization as efficient route for improvement of lignin flame retardant action in PLA [J]. European Polymer Journal, 84: 652-667.

DABABI I, GIMELLO O, ELALOUI E, et al. 2016. Organosolv lignin based wood adhesive. Influence of the lignin extraction conditions on the adhesive performance [J]. Polymers, 8 (9): 340.

De CHIRICO A, ARMANINI M, CHINI P, et al. 2003. Flame retardants for polypropylene based on lignin [J]. Polymer Degradation and Stability, 79 (1): 139-145.

De LOS SANTOS RAMOS W, POZNYAK T, CHAIREZ I, et al. 2009. Remediation of lignin and its derivatives from pulp and paper industry wastewater by the combination of chemical precipitation and ozonation [J]. Journal of Hazardous Materials, 169 (1-3): 428-434.

DEAN A P, SIGEE D C, ESTRADA B, et al. 2010. Using FTIR spectroscopy for rapid determination of lipid accumulation in response to nitrogen limitation in freshwater microalgae [J]. Bioresource Technology, 101 (12): 4499-4507.

EL MANSOURI N, SALVADO J. 2007. Analytical methods for determining functional groups in various technical lignins [J]. Industrial Crops and Products, 26 (2): 116-124.

ERDOCIA X, PRADO R, ANGELES CORCUERA M, et al. 2014. Effect of different organosolv treatments on the structure and properties of olive tree pruning lignin [J]. Journal of Industrial and Engineering Chemistry, 20 (3): 1103-1108.

FAIX O. 1991. Classification of lignins from different botanical origins by FTIR spectroscopy [J]. Holzforschung, 45 (s1): 21-27.

FAIX O, ANDERSONS B, ZAKIS G. 1998. Determination of carbonyl groups of six round robin lignins by modified oximation and FTIR spectroscopy [J]. Holzforschung, 52 (3): 268-274.

FAUSTINO H, GIL N, BAPTISTA C, et al. 2010. Antioxidant activity of lignin phenolic compounds extracted from kraft and sulphite black liquors [J]. Molecules, 15 (12): 9308-9322.

FELDMAN D, BANU D, EL-AGHOURY A. 2007. Plasticization effect of lignin in some highly filled vinyl formulations [J]. Journal of Vinyl and Additive Technology, 13 (1): 14-21.

FERNANDES E M, AROSO I M, MANO J F, et al. 2014. Functionalized cork-polymer composites (CPC) by reactive extrusion using suberin and lignin from cork as coupling agents [J]. Composites Part B-Engineering, 67: 371-380.

FU R L, CHENG X S. 2011. Synthesis and flame retardant properties of melamine modified EH lignin [J]. Advanced Materials Research, 236-238: 482-485.

GE Y, QIN L, LI Z. 2016. Lignin microspheres: An effective and recyclable natural polymer-based adsorbent for lead ion removal [J]. Materials and Design, 95: 141-147.

GUO X, WU Y, YAN N. 2017. Characterizing spatial distribution of the adsorbed water in wood cell wall of Ginkgo biloba L. by μ-FTIR and confocal Raman spectroscopy [J]. Holzforschung, 71 (5):

415-423.

GUO X, WU Y, YAN N. 2018. *In situ* micro-FTIR observation of molecular association of adsorbed water with heat-treated wood [J]. Wood Science and Technology, 52 (4): 971-985.

HU S, HSIEH Y. 2016. Silver nanoparticle synthesis using lignin as reducing and capping agents: A kinetic and mechanistic study [J]. International Journal of Biological Macromolecules, 82: 856-862.

KAI D, REN W, TIAN L, et al. 2016. Engineering poly(lactide)-lignin nanofibers with antioxidant activity for biomedical application [J]. ACS SUSTAINABLE CHEMISTRY and ENGINEERING, 4 (10): 5268-5276.

KASPRZYCKA-GUTTMAN T, ODZENIAK D. 1994. Antioxidant properties of lignin and its fractions [J]. Thermochimica Acta, 231: 161-168.

KHAN M A, ASHRAF S M, MALHOTRA V P. 2004. Development and characterization of a wood adhesive using bagasse lignin [J]. International Journal of Adhesion and Adhesives, 24 (6): 485-493.

KILPELÄINEN I, ÄMMÄLAHTI E, BRUNOW G, et al. 1994. Application of three-dimensional HMQC-HOHAHA NMR spectroscopy to wood lignin, a natural polymer [J]. Tetrahedron Letters, 35 (49): 9267-9270.

KOUMBA-YOYA G, STEVANOVIC T. 2017. Study of organosolv lignins as adhesives in wood panel production [J]. Polymers, 9 (2): 46.

LIU Y, YANG Z, DESYATERIK Y, et al. 2008. Hygroscopic behavior of substrate-deposited particles studied by micro-FTIR spectroscopy and complementary methods of particle analysis [J]. Analytical Chemistry, 80 (3): 633-642.

MENG X, SEDMAN J, van de VOORT F R. 2012. Improving the determination of moisture in edible oils by FTIR spectroscopy using acetonitrile extraction [J]. Food Chemistry, 135 (2): 722-729.

MILCZAREK G. 2010. Kraft lignin as dispersing agent for carbon nanotubes [J]. Journal of Electroanalytical Chemistry, 638 (1): 178-181.

MOREIRA J L, SANTOS L. 2004. Spectroscopic interferences in Fourier transform infrared wine analysis [J]. Analytica Chimica Acta, 513 (1): 263-268.

MURAILLE L, PERNES M, HABRANT A, et al. 2015. Impact of lignin on water sorption properties of bioinspired self-assemblies of lignocellulosic polymers [J]. European Polymer Journal, 64: 21-35.

NAIR S S, KUO P, CHEN H, et al. 2017. Investigating the effect of lignin on the mechanical, thermal, and barrier properties of cellulose nanofibril reinforced epoxy composite [J]. Industrial Crops and Products, 100: 208-217.

NITZ H, SEMKE H, MULHAUPT R. 2001. Influence of lignin type on the mechanical properties of lignin based compounds [J]. Macromolecular Materials and Engineering, 286 (12): 737-743.

OLSSON A M, SALMEN L. 2004. The association of water to cellulose and hemicellulose in paper examined by FTIR spectroscopy [J]. Carbohydrate Research, 339 (4): 813-818.

PASSAUER L, STRUCH M, SCHULDT S, et al. 2012. Dynamic moisture sorption characteristics of xerogels from water swellable oligo(oxyethylene) lignin derivatives [J]. Acs Applied Materials and Interfaces, 4 (11): 5852-5862.

PEPPER J M, BAYLIS P E T, ADLER E. 1959. The isolation and properties of lignins obtained by the acidolysis of spruce and aspen woods in dioxane–water medium [J]. Canadian Journal of Chemistry, 37 (8): 1241-1248.

POPESCU C, POPESCU M, VASILE C. 2010. Structural changes in biodegraded lime wood [J]. Carbohydrate Polymers, 79 (2): 362-372.

POUTEAU C, DOLE P, CATHALA B, et al. 2003. Antioxidant properties of lignin in polypropylene [J]. Polymer Degradation and Stability, 81 (1): 9-18.

RAHMAN M M, ZAHIR M H, KIM H D. 2016. Synthesis and properties of waterborne polyurethane (WBPU)/modified lignin amine (MLA) adhesive: a promising adhesive material [J]. Polymers, 8 (9): 318.

RAWAT S P S, KHALI D P. 1999. Studies on adsorption behavior of water vapor in lignin using the Brunauer-Emmett-Teller theory [J]. Holz Als Roh- Und Werkstoff, 57 (3): 203-204.

REINA J J, DOMINGUEZ E, HEREDIA A. 2001. Water sorption-desorption in conifer cuticles: the role of lignin [J]. Physiologia Plantarum, 112 (3): 372-378.

ROZMAN H D, TAN K W, KUMAR R N, et al. 2001. Preliminary studies on the use of modified ALCELL lignin as a coupling agent in the biofiber composites [J]. Journal of Applied Polymer Science, 81 (6): 1333-1340.

SADEGHIFAR H, ARGYROPOULOS D S. 2015. Correlations of the antioxidant properties of softwood kraft lignin fractions with the thermal stability of its blends with polyethylene [J]. ACS Sustainable Chemistry and Engineering, 3 (2): 349-356.

SHAHA S K, DYUTI S, AHSAN Q et al. 2011. Effect of alkali treatment on surface morphology and properties of jute yarns [J]. Advanced Materials Research, 264-265: 1922-1927.

SONG P, CAO Z, FU S, et al. 2011. Thermal degradation and flame retardancy properties of ABS/lignin: effects of lignin content and reactive compatibilization [J]. Thermochimica Acta, 518 (1-2): 59-65.

SONG Y, WANG Z, YAN N, et al. 2016. Demethylation of wheat straw alkali lignin for application in phenol formaldehyde adhesives [J]. Polymers, 8 (6): 209.

SPIRIDON I, LELUK K, RESMERITA A M, et al. 2015. Evaluation of PLA-lignin bioplastics properties before and after accelerated weathering [J]. Composites Part B-Engineering, 69: 342-349.

STEWART D. 2008. Lignin as a base material for materials applications: Chemistry, application and economics [J]. Industrial Crops and Products, 27 (2): 202-207.

STIUBIANU G, NISTOR A, VLAD A, et al. 2011. Modification of water sorption capacity of polydimethylsiloxane based composites by incorporation of lignin [J]. Materiale Plastice, 48 (4): 289-294.

THAKUR V K, THAKUR M K. 2015. Recent advances in green hydrogels from lignin: a review [J]. International Journal of Biological Macromolecules, 72: 834-847.

VOLKOVA N, IBRAHIM V, HATTI-KAUL R, et al. 2012. Water sorption isotherms of Kraft lignin and its composites [J]. Carbohydrate Polymers, 87 (2): 1817-1821.

VU T, CHAFFEE A, YAROVSKY I. 2002. Investigation of lignin-water interactions by molecular simulation [J]. Molecular Simulation, 28 (10-11): 981-991.

WARTH H, MÜLHAUPT R, SCHÄTZLE J. 1997. Thermoplastic cellulose acetate and cellulose acetate compounds prepared by reactive processing [J]. Journal of Applied Polymer Science, 64 (2): 231-242.

XIONG F, WU Y, LI G, et al. 2018. Transparent nanocomposite films of lignin nanospheres and poly(vinyl alcohol) for UV-absorbing [J]. Industrial and Engineering Chemistry Research, 57 (4): 1207-1212.

XU F, YU J, TESSO T, et al. 2013. Qualitative and quantitative analysis of lignocellulosic biomass using infrared techniques: a mini-review [J]. Applied Energy, 104: 801-809.

XU G, YAN G, ZHANG J. 2015. Lignin as coupling agent in EPDM rubber: Thermal and mechanical properties [J]. Polymer Bulletin, 72 (9): 2389-2398.

YANEZ-S M, MATSUHIRO B, NUNEZ C, et al. 2014. Physicochemical characterization of ethanol organosolv lignin (EOL) from Eucalyptus globulus: effect of extraction conditions on the molecular structure [J]. Polymer Degradation and Stability, 110: 184-194.

YOSHIDA H, MÖRCK R, KRINGSTAD K P, et al. 1987. Kraft lignin in polyurethanes I. mechanical properties of polyurethanes from a Kraft lignin-polyether triol-polymeric MDI system [J]. Journal of Applied Polymer Science, 34 (3): 1187-1198.

YU Y, FU S, SONG P, et al. 2012. Functionalized lignin by grafting phosphorus-nitrogen improves the thermal stability and flame retardancy of polypropylene [J]. Polymer Degradation and Stability, 97 (4): 541-546.

ZHU H, CHEN Y, QIN T, et al. 2014. Lignin depolymerization via an integrated approach of anode oxidation and electro-generated H_2O_2 oxidation [J]. RSC Advances, 4 (12): 6232-6238.

Chapter 4 Quantitatively characterizing moisture sorption of cellulose using micro-FTIR spectroscopy

4.1 Introduction

Since the beginning of the 21st century, non- renewable resources such as natural gas, oil, coal and other fossil resources have dried up, and a large number of fossil resources have lead to air pollution, global warming and other environmental problems. Biomass resource has been paid more and more attention because of its advantages of wide source, environmental protection and renewable. The largest reserves of biomass resources are plant fibers. More than 50 billion tons of plant fiber are grown on land each year, accounting for nearly 60 to 80 percent of the planet's biomass. Cellulose is the main component of plant fiber, which is the most abundant and most widely distributed natural renewable resources in nature. Cotton contains nearly 100% cellulose, hemp about 80% to 90% cellulose, and wood and bamboo about 40% to 50% cellulose. In addition, cellulose can also be taken from bacteria, bagasse, straw, seaweed and so on. Our country is a country with vast land and abundant resources. The raw materials used for cellulose extraction are very rich (Berglund et al. 2016; De Rosa et al. 2010; Hassan et al. 2012; Ruan et al. 1996). Cellulose has long been used in the apparel and paper industries for its advantages of low cost, high stiffness and low density. The sustainable development of economy has become the theme of every country. As the most abundant natural organic macromolecule, cellulose has great value in economic development and environmental protection.

With the demand for cellulose products, cellulose nanofiber film has emerged. Cellulose nanofiber film, as a novel cellulose derivation, has excellent physical, chemical, and biological properties, such as good thermal stability, high strength, low degradation, and non-toxicity (Gamelas et al. 2015; Liu et al. 2016; Sun et al. 2015; Xu et al. 2016), and it is a very promising candidate for tissue engineering, electronics, and green packaging materials (Deng et al. 2017; Khalil et al. 2012; Khalil et al. 2016;

Manhas et al. 2015; Shi et al. 2014; Stelte and Sanadi 2009). However, hydrophilic cellulose nanofiber film adsorbs water under hydrothermal conditions, which strongly affects its surface behavior and can lead to reliability problems (Isa et al. 2013; Uraki et al. 2010; Vogt et al. 2005). Consequently, the water adsorption mechanisms of cellulose nanofiber film need to be fully understood in order for this sustainable raw material to be efficiently utilized.

Cellulose materials are completely degradable and hygroscopic. As a wet core, it has been used in intermittent commutative heat exchanger, and its influence on the performance of heat exchanger has been studied experimentally. At present, the hygroscopic isotherms of fiber materials are obtained by experiments, and the process of hygroscopicity and dehumidification of fiber are described, but the quantitative method needs to be determined. Water adsorption is a key property of cellulose materials (Angkuratipakorn et al. 2017; Peresin et al. 2010; Wan et al. 2009; Watanabe et al. 2006), and it has been studied via a number of experimental approaches, such as quartz crystal microgravimetry (Kittle et al. 2011), dynamic vapor sorption (Agrawal et al. 2004; Driemeier et al. 2012; Hill et al. 2010; Popescu et al. 2014; Rautkari et al. 2013; Zaihan et al. 2009), dielectric relaxation spectroscopy (Smith 1995; Sugimoto et al. 2008), Fourier transform infrared spectroscopy (Célino et al. 2014; Murphy and Depinho 1995; Olsson and Salmen 2004), and nuclear magnetic resonance spectroscopy (Bergenstrahle et al. 2008; Carles and Scallan 1973; Daniel and Olle 2001; Felby et al. 2008; Hall and Rajanayagam 1986; Menon et al. 1987; Ogiwara et al. 1969). Of these experimental approaches, dynamic vapor sorption (DVS) is an important quantitative method which has been widely used and has provided vast amounts of water sorption data (Glass et al. 2018; Hill et al. 2012). Using this method, the sorption isotherm and sorption hysteresis of many cellulose materials including natural fibers (Hill et al. 2010; Kohler et al. 2003; Xie et al. 2011), regenerated cellulose (Okubayashi et al. 2005), microcrystalline cellulosic fibers (Kachrimanis et al. 2006; Xie et al. 2011), and wood powder (Madamba et al. 1996), have been analyzed. As this method has been confirmed to be able to give highly reproducible sorption data of tested sample over a wide RH range in real time (Hill et al. 2010; Lundahl et al. 2016), it can be used as a reference method to quantitatively analyze the water adsorption. Meanwhile, Fourier transform infrared (FTIR) spectroscopy shows some distinct advantages because its spectral information allowed both qualitative and quantitative analyses of the water adsorption (Guo et al. 2018). In addition, micro-FTIR spectroscopy has recently been developed, and it has better sensitivity for the detection of water than conventional FTIR approaches. This is because micro-FTIR spectroscopy equipment contains an additional visible-light microscope that can be used to visualize the morphology and select observation area of micro-sized sample. It has become a powerful tool for studying the chemical composition and spatial distribution of complex substances (Loutherback et al. 2016; Mazzeo and Joseph 2007). As a result, micro-FTIR spectroscopy is believed to hold promise for the investigation of the water adsorption of nanocellulose materials such as cellulose nanofiber film.

Many interesting results about the water adsorption characteristics of cellulose materials have been obtained using FTIR spectroscopy (Dias et al. 1998; Fengel 1993; Gu et al. 2004; Hou et al. 2014). These works given as examples in this paragraph are very important, as they have provided much FTIR spectral information for qualitatively characterizing water adsorption of cellulose materials. However, based on univariate analysis of this FTIR spectral information, water sorption isotherms of cellulose materials were not properly determined.

The aim is the use of micro-FTIR spectroscopy as an experimental tool to qualitatively and quantitatively characterize water adsorption of nanocellulose materials (i.e., cellulose nanofiber film). Firstly, we collected *in situ* micro-FTIR spectra of cellulose nanofiber film over a wide range of RH levels to indicate the effective adsorption sites for adsorbed water and identify the spectral regions affected by water adsorption. We also measured moisture contents of cellulose nanofiber film during water sorption process using DVS apparatus as reference values. Secondly, a multivariate model linking the measured moisture contents and the selected micro-FTIR spectral regions was developed using partial least square regression (PLSR) method. Finally, we estimated sorption isotherms using this constructed multivariate model, and the FTIR spectra were used to quantitatively characterize the water absorption of cellulose nanofiber film at 25 ℃.

4.2 Experiment section

4.2.1 Sample preparation

This book focused on a typical specimen of nanocellulose material (i.e., cellulose nanofiber film), and the preparation of this cellulose nanofiber film was described clearly in earlier reports (Han et al. 2013). In the present study, we used the same procedure. Once the film was obtained, it was covered by two pieces of cover slips and dried under atmospheric conditions. Before the spectral experiment, the film was dried a further time in nitrogen gas for at least 2 h.

4.2.2 Micro-FTIR spectroscopy apparatus

Figure 4.1 shows a diagram of the micro-FTIR spectroscopy apparatus. The main part of the apparatus was a micro-FTIR spectrometer (Nicolet IN 10^{TM}, Thermo Electron Scientific Instruments, Madison, WI, USA). This spectrometer was equipped with one more microscope than was found in conventional FTIR spectrometers; in this microscope, two pathways for visible and IR beams were included; the visible beam came from a visible light source, and it could be used for visual examinations and the selection of an observation area. This microscope was set to the optical mode, and the aperture was adjusted so as to be 100 μm by 100 μm; the desired observation areas for the cellulose nanofiber film were then defined. The micro-FTIR spectrum was collected using 64 scans between 720-4000 cm^{-1} using 4 cm^{-1} spectral resolutions. In order to obtain a higher signal-to-noise ratio, the spectra for cellulose nanofiber film and background were collected one after another in order to compensate for any moisture that may exist in the sample chamber.

Figure 4.1 also shows a specially designed sample chamber. Once the cellulose nanofiber film was placed on the bottom of the chamber (which was composed of a ZnSe plate), the sample chamber was sealed by a lid (which contained another ZnSe

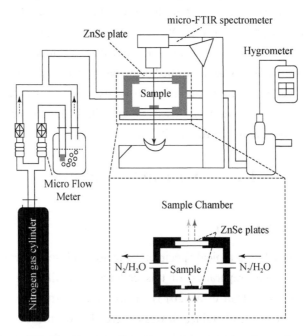

Figure 4.1 The diagram of experimental apparatus used to study water adsorption of cellulose nanofiber film

plate). The ZnSe plates provided the optical paths for both the visible and IR light through this sample chamber. The RH in this chamber was accurately controlled using a stream of dry nitrogen and saturated vapor, whose ratio was changed using a high-precision gas micro-flow meter (Alicat Scientific, Tucson, AZ, USA). The resulting RH was then measured using a humidity/temperature meter (Center 310, Center Technology Corp., New Taipei, Taiwan, China).

In the experiment, even if the RH of the chamber reaches the set value, a period of 60 min should be maintained. The equilibrium time was from a kinetic spectroscopic test. Before the start of this test, cellulose nanofiber film was equilibrated at the starting RH, i.e., 0%RH, for 4 h. We took the first spectrum measurement after starting, after which we changed the RH of the sample chamber to 5% and then recorded the spectra every 0.5 min. As shown in Figure 4.2, after changing the target RH to a new setting, i.e., 5%RH, there was typically a delay of approximately 3-4 min in which the actual RH approached the set target RH. While, there were no changed detected in the measured spectra after a period of 15 min, and this situation remained the after 100 min (the peak height of the major peak at 3352 cm^{-1} was used to demonstrate the spectral change). And then the target RH was changed to next setting, i.e., 10%RH, no detected in the measured spectra after a period of 15 min. In addition, the temperature in the specially designed sample chamber was maintained at 25°C. Based on these kinetic test results, we maintained a conditioning time of 60 min at each value prior to measurement allow the sample to stabilize at the specific RH.

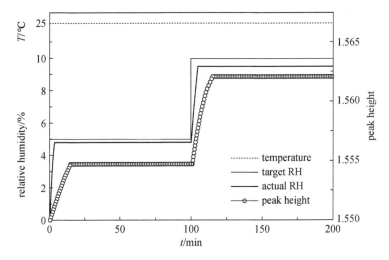

Figure 4.2 Typical changes of target RH, actual RH and peak height of the major peak at 3352 cm^{-1} during the water adsorption process. The temperature was maintained at 25 °C

4.2.3 DVS apparatus

A DVS apparatus (DVS AdvantagePlus, Surface Measurement Systems Ltd, London, U.K.) was used to determine the moisture content of cellulose nanofiber film. Data such as the run time, real-time masses of the samples, and RH at a constant temperature of 25°C were obtained during the water sorption process. The RH was set to change from 0% to 95% in 5% steps and then decreased to 0%. At every stage, the RH was kept constant for some time and then increased to the next increment as the sample mass decreased at 0.002% min^{-1} over a 10 min period. The moisture content was calculated using real-time mass measurements of the sample and the following equation:

$$\mathrm{MC} = \frac{m_2 - m_1}{m_1} \times 100\% \tag{4.1}$$

Where, MC is the moisture content of the sample, m_1 is the dry mass of the sample, and m_2 is the real-time mass of the sample at a set RH.

In the experiment, typical changes of these data were shown in Figure 4.3. After changing the target RH to a new setting, there was a delay of approximately 4-10 min in which the actual RH approached the set target RH. After the initial few minutes, the actual RH was kept stable, and then the fluctuation of the actual RH value was less than 0.1% at extended time. The real-time moisture content of the sample generated an asymptotic curve against time, and then reached the equilibrium moisture content. Meanwhile, it was found that the temperature values were very stable.

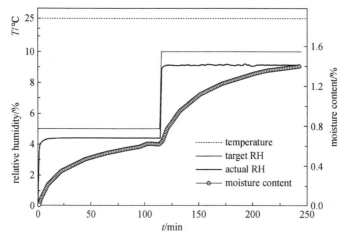

Figure 4.3 Typical changes of target RH, actual RH and moisture content during the water adsorption process. The temperature was maintained at 25°C

4.2.4 Spectral data processing

1. Obtaining difference spectra

In order to clearly demonstrate minor spectral changes during water adsorption process, difference spectra were obtained in the OMINIC 8.0 software by FTIR subtractive spectroscopy technique. By using this technique, difference spectra were acquired by subtracting the spectra at different RH values from the spectrum of dry cellulose nanofiber film in a 1 ∶ 1 ratio.

2. Building quantitative prediction model

The quantitative prediction model based on micro-FTIR spectra was established by PLSR method. In the process of establishing models, the micro-FTIR spectra of the cellulose nanofiber film in the RH range from 0% to 94% were all imported into TQ Analyst 9 (Thermo Scientific Inc., Madison, Wisconsin, USA). The type of optical path was set to "constant". The moisture content was defined as the only measured component, and the upper and lower values of the moisture content were set to 28 % and 0 %, and the measured moisture contents using DVS apparatus were imported as reference values. No filters or no baseline corrections were used in the spectral preprocessing, and some regions of the original spectra associated with water adsorption were selected to construct models. Moreover, in order to avoid bias in the subset, the original spectra of cellulose nanofiber film at five replicates for each of 12 RH levels (i.e., 60 spectra) were randomly selected into calibration groups, while five replicates per four other RH levels (i.e., 20 spectra) were selected as a validation group. The randomly selected spectra in the calibration group were used to perform calculations and cross-validations; for this procedure, the best predictive model with the highest correlation of determination (R^2) and lowest root mean square error of cross validation (RMSECV) was regressed after several iterations. Meanwhile, the spectra in the validation group were used for model evaluation. For this procedure, a predictive model was determined to be one with a high correlation of determination (R^2) and a low root mean square error of prediction (RMSEP).

4.3　Results and discussion

4.3.1　Qualitatively analyzing water adsorption of cellulose nanofiber film

Figure 4.4 shows the micro-FTIR spectra of cellulose nanofiber film at different RH levels during the water adsorption process. Meanwhile, the assignments of the main spectral peaks associated with water adsorption in the micro-FTIR spectrum of cellulose nanofiber film are summarized in Table 4.1 (Abidi et al. 2014; Liang and Marchessault 1959; Rosa et al. 2010).

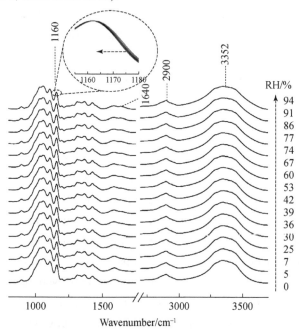

Figure 4.4   Micro-FTIR spectra of cellulose nanofiber film during the water adsorption process. The arrow demonstrated that the RH values were increased from 0% to 94%

Table 4.1   Assignment of the main absorption bands in micro-FTIR spectra of cellulose nanofiber film which were associated with water adsorption

Wavenumber/cm^{-1}	Assignment
3700-3000	Hydrogen bonded of O—H stretching vibration
3000-2700	C—H stretching vibration
1640	O—H bending vibration of adsorbed water
1163	C—O—C asymmetric stretching at β-glucosidic linkage

As shown in Figure 4.4, the most intense peak at 3352 cm^{-1} was attributed to O—H stretching vibration of cellulose nanofiber film and absorbed water, and it increased with a rise in RH. The weaker peak at 1640 cm^{-1} was assigned to the O—H bending vibration of adsorbed water, and it was also observed to increase as the RH increased from 0 to 94%. The spectral changes of these two peaks confirmed that OH groups were effective adsorption sites for adsorbed water. Moreover, the peak at 1160 cm^{-1} was assigned to that of C—O—C asymmetric stretching vibration at the β-glucosidic linkage, which was related to the cellulose chain. A continuous shift of this peak to lower wave numbers was observed as the RH increased. This shift was also caused by water adsorption, for absorbed water could stiffen the cellulose chain. Based on these characteristics, it was found that the spectral regions affected by the water adsorption were 3700-3000 cm^{-1}, 1740-1618 cm^{-1}, and 1190-1139 cm^{-1}.

Water adsorption in micro-FTIR spectroscopy provided unresolved broad peaks, which offered limited precise information. With the purpose of extracting minor spectral changes, difference spectra were obtained. Figure 4.5 shows difference spectra of cellulose nanofiber film at various RH levels from 5% to 94%. The signature of the main peak in the spectral region of 3700-2852 cm^{-1} increased which was identified and assigned to absorbed water. Similar growth trends of this broad peak had been found in other measurements of cellulose materials (Guo et al. 2016). It can be concluded that the whole spectral region of 3700-2852 cm^{-1} was related to the water adsorption. Meanwhile, the positive-going peak at 1646 cm^{-1} changed continuously whilst the RH increased. The intensity of the peak at 1176 cm^{-1} was observed to decrease with an increase in RH, while the signature from the peak at 1157 cm^{-1} was observed to rise. The changes of these three peaks further indicated that the spectral regions of 1740-1618 cm^{-1}, and 1190-1139 cm^{-1} were affected by the water adsorption. In summary, the study results demonstrated that the regions of micro-FTIR spectra closely associated with water adsorption were 3700-2852 cm^{-1}, 1740-1618 cm^{-1}, and 1190-1139 cm^{-1}.

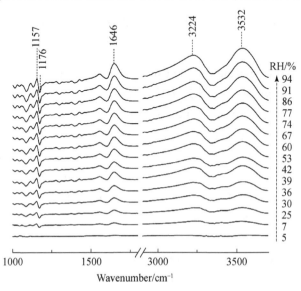

Figure 4.5　Difference micro-FTIR spectra of cellulose nanofiber film during the water adsorption process. The arrow demonstrated that the RH values were increased from 5% to 94%

Figure 4.6 shows the development of the peak height as a function of RH, for these three major peaks affected by the water adsorption. As shown in this figure, these three major peaks had different growth rates. The discrepancies were observed in many cellulose materials. Moreover, a univariate analysis of these peaks failed to properly reproduce the sorption isotherms. Therefore, a multivariate approach, such as a PLSR method, should be introduced to qualitatively characterize the water adsorption.

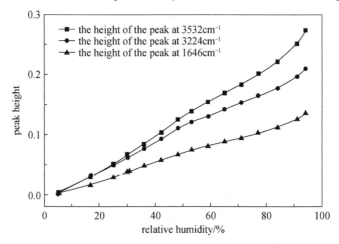

Figure 4.6 Peak height of three major peaks associated with water adsorption taken from difference spectra over the set RH range

4.3.2 Quantitative analysis of water adsorption in cellulose nanofiber film

1. Obtaining the reference value using DVS apparatus

As mentioned in Section 4.1, DVS can be used as a reference method to quantitatively analyze water adsorption. Here, using this apparatus, moisture contents of cellulose nanofiber film were obtained as reference values. Figure 4.7(a) showed the moisture content of cellulose nanofiber film with the varying RH levels over the time profile in the isotherm run. In this figure, the RH was changed step-by-step and the sorption kinetics was recorded. Based on this, by plotting the respective equilibrium moisture contents versus the corresponding level of RH, sorption isotherm was obtained as shown in Figure 4.7(b).

2. Development of the PLSR quantitative models

As mentioned in Section 4.2.4, the spectral regions are important parameters for constructing a PLSR quantitative model. Here, three spectral regions of 3700-2852 cm^{-1}, 1740-1618 cm^{-1}, and 1190-1139 cm^{-1} affected by water adsorption were used to

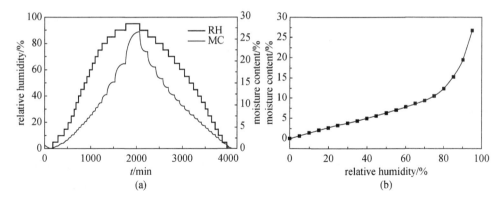

Figure 4.7 (a) The moisture content of cellulose nanofiber film with the varying RH levels over the time profile in the isotherm run; (b) The sorption isotherm of cellulose nanofiber film measured by DVS apparatus

construct the quantitative prediction model as one proposed case. On this basis, the increase and decrease of these three spectral regions were also introduced as the other two cases for comparison (3700-2852 cm^{-1}, 2852-2538 cm^{-1}, 1740-1618 cm^{-1}, and 1190-1139 cm^{-1}, second case; 3700-3000 cm^{-1}, 2852-2538 cm^{-1}, 1740-1618 cm^{-1}, and 1190-1139 cm^{-1}, third case). These three cases and their corresponding spectral regions, RESECV, RESEP, and R^2 values in the PLSR models are all summarized in Table 4.2. These PLSR quality parameters such as RESECV, RESEP, and R^2 were described before (Fitzpatrick et al. 2012). The first model using three spectral regions of 3700-2852 cm^{-1}, 1740-1618 cm^{-1}, and 1190-1139 cm^{-1} had the highest coefficients of determination, lowest RMSEP and RMSECV values which showed that this model had good estimated performance and high prediction accuracy. What's more, these results suggested that the best model was one which used complete spectral regions closely associated with water adsorption, and the increase or decrease of these spectral regions would lower the quality of the quantitative model.

Table 4.2 PLSR quality parameters for cross and test set validation for the proposed three cases

	First case	Second case	Third case
Spectral region/cm^{-1}	3700-3000/1740-1618/ 1190-1139	3700-2852/1740-1618/ 1190-1139	3700-2852/2852-2538/ 1740-1618/1190-1139
Cross validation			
RMSECV	0.658	0.637	0.882
Number of PLS components	6	6	6
R^2/%	99.62	99.66	99.35
External validation			

	First case	Second case	Third case
RMSEP	0.530	0.439	0.561
Number of PLS components	6	6	6
R^2/%	98.94	99.52	99.29

RMSECV, root mean square error of cross validation; RMSEP, root mean square error of prediction; R^2, coefficient of determination.

With the help of this optimal PLSR quantitative model, we estimated the moisture contents of cellulose nanofiber film using the micro-FTIR spectra collected over the entire RH range of 0%-94%. The calculated moisture contents of cellulose nanofiber film in the full set RH range during the adsorption process in the isothermal sorption run are shown in Figure 4.8. This sorption isotherm exhibited a typical sigmoidal-shaped curve in the adsorption isotherm run; this has been observed in other cellulose materials(Alix et al. 2009; Gouanve et al. 2007). The water adsorption process of cellulose nanofiber film can also be separated into three stages, as explained in previous studies. In the first stage (RH ⩽ 35%), the absorbed water is bound directly to the chemical adsorption sites through hydrogen bonding, and related to Langmuir's mode. In the second stage (35% < RH < 55%), the absorbed water is indirectly bonded to hydrophilic groups through the absorbed water, and the water content increases with RH as per Henry's model. In the third section (RH ⩾ 55%), the absorbed water interacts with them, and forms a five-molecule tetrahedral structure. As shown in Figure 4.8, the predicted values and reference values were very similar (relative error

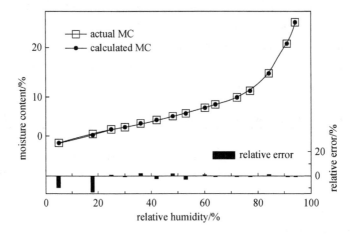

Figure 4.8 Comparison of the sorption isotherms of cellulose nanofiber film calculated by the PLSR quantitative model and measured by DVS approaches

was lower than 3%), and it confirmed the effectiveness of micro-FTIR spectroscopy in characterizing water adsorption of cellulose nanofiber film. Meanwhile, the relative errors were different in different stages of the water adsorption process, and this difference may be due to above-mentioned adsorption mechanism.

4.4 Conclusions

By using *in situ* micro-FTIR spectroscopy, a detailed investigation on water adsorption of cellulose nanofiber film was achieved. As the RH increased from 0% to 94%, the increase of the characteristic peaks at 3352 cm^{-1} and 1604 cm^{-1} indicated that OH groups were effective adsorption sites for adsorbed water. Meanwhile, the difference between the moist and dry spectra identified that three spectral regions of 3700-2852 cm^{-1}, 1740-1618 cm^{-1}, and 1190-1139 cm^{-1} were closely related to the water adsorption. On this basis, using partial least square regression (PLSR), three multivariate models linking moisture contents and the identified micro-FTIR spectral regions were developed. By comparing these models, it was found that the best model was one which used complete spectral regions closely associated with water adsorption. Furthermore, using this optimal PLSR quantitative model, water sorption isotherm was determined, and it showed that the predicted moisture contents were well consistent with the reference values obtained using DVS apparatus in the RH region of 35%-94% (relative error was lower than 3%). This study confirmed that the effectiveness of micro-FTIR spectroscopy in characterizing water adsorption of cellulose nanofiber film at room temperature of 25℃, which could assist in developing a non-destructive and rapid method for qualitatively and quantitatively analyzing water adsorption of nanocellulose materials.

References

ABIDI N, CABRALES L, HAIGLER C H. 2014. Changes in the cell wall and cellulose content of developing cotton fibers investigated by FTIR spectroscopy [J]. Carbohydrate Polymers, 100 (SI): 9-16.

AGRAWAL A M, MANEK R V, KOLLING W M, et al. 2004. Water distribution studies within microcrystalline cellulose and chitosan using differential scanning calorimetry and dynamic vapor sorption analysis [J]. Journal of Pharmaceutical Sciences, 93 (7): 1766-1779.

ALIX S, PHILIPPE E, BESSADOK A, et al. 2009. Effect of chemical treatments on water sorption and mechanical properties of flax fibers [J]. Bioresource Technology, 100 (20): 4742-4749.

ANGKURATIPAKORN T, SINGKHONRAT J, CHRISTY A A. 2017. Comparison of water adsorption properties of cellulose and cellulose nanocrystals studied by near-infrared spectroscopy and gravimetry [J]. Key Engineering Materials, 735: 235-239.

BERGENSTRAHLE M, WOHLERT J, LARSSON P T, et al. 2008. Dynamics of cellulose-water interfaces: NMR spin-lattice relaxation times calculated from atomistic computer simulations [J]. Journal of Physical Chemistry B, 112 (9): 2590-2595.

BERGLUND L, NOEL M, AITOMAKI Y, et al. 2016. Production potential of cellulose nanofibers from industrial residues: Efficiency and nanofiber characteristics [J]. Industrial Crops and Products, 92: 84-92.

CARLES J E, SCALLAN A M. 1973. Determination of amount of bound water within cellulosic gels by nmr-spectroscopy [J]. Journal of Applied Polymer Science, 17 (6): 1855-1865.

CÉLINO A, GONCALVES O, JACQUEMIN F, et al. 2014. Qualitative and quantitative assessment of water sorption in natural fibers using ATR-FTIR spectroscopy [J]. Carbohydrate Polymers, 101: 163-170.

DANIEL T, OLLE S. 2001. Diffusion of water absorbed in cellulose fibers studied with ^{1}H-NMR [J]. Langmuir, 17 (9): 2694-2702.

De ROSA I M, KENNY J M, PUGLIA D, et al. 2010. Morphological, thermal and mechanical characterization of okra (Abelmoschus esculentus) fibers as potential reinforcement in polymer composites [J]. Composites Science and Technology, 70 (1): 116-122.

DENG Z, JUNG J, ZHAO Y. 2017. Development, characterization, and validation of chitosan adsorbed cellulose nanofiber (CNF) films as water resistant and antibacterial food contact packaging [J]. Lwt-Food Science and Technology, 83: 132-140.

DIAS C R, ROSA M J, de PINHO M N. 1998. Structure of water in asymmetric cellulose ester membranes — and ATR-FTIR study [J]. Journal of Membrane Science, 138 (2): 259-267.

DRIEMEIER C, MENDES F M, OLIVEIRA M M. 2012. Dynamic vapor sorption and thermoporometry to probe water in celluloses [J]. Cellulose, 19 (4): 1051-1063.

FELBY C, THYGESEN L G, KRISTENSEN J B, et al. 2008. Cellulose-water interactions during enzymatic hydrolysis as studied by time domain NMR [J]. Cellulose, 15 (5): 703-710.

FENGEL D. 1993. Influence of water on the oh valency range in deconvoluted ftir spectra of cellulose [J]. Holzforschung, 47 (2): 103-108.

FITZPATRICK M, CHAMPAGNE P, CUNNINGHAM M F. 2012. Quantitative determination of cellulose dissolved in 1-ethyl-3-methylimidazolium acetate using partial least squares regression on FTIR spectra [J]. Carbohydrate Polymers, 87 (2): 1124-1130.

GAMELAS J A F, PEDROSA J, LOURENCO A F, et al. 2015. On the morphology of cellulose nanofibrils obtained by TEMPO-mediated oxidation and mechanical treatment [J]. Micron, 72: 28-33.

GLASS S V, BOARDMAN C R, THYBRING E E, et al. 2018. Quantifying and reducing errors in equilibrium moisture content measurements with dynamic vapor sorption (DVS) experiments [J]. Wood Science and Technology, 52 (4): 909-927.

GOUANVE F, MARAIS S, BESSADOK A, et al. 2007. Kinetics of water sorption in flax and PET fibers [J]. European Polymer Journal, 43 (2): 586-598.

GU J, GAO Z, LI Z, et al. 2004. The FTIR study on the reaction of benzyl isocyanate and cellulose with different moisture content [J]. Scientia Silvae Sinicae, 40 (2): 142-147.

GUO X, LIU L, WU J, et al. 2018. Qualitatively and quantitatively characterizing water adsorption of a cellulose nanofiber film using micro-FTIR spectroscopy [J]. RSC Advances, 8 (8): 4214-4220.

GUO X, QING Y, WU Y, et al. 2016. Molecular association of adsorbed water with lignocellulosic materials examined by micro-FTIR spectroscopy [J]. International Journal of Biological Macromolecules, 83: 117-125.

HALL L D, RAJANAYAGAM V. 1986. Evaluation of the distribution of water in wood by use of 3-dimensional proton NMR volume imaging [J]. Wood Science and Technology, 20 (4): 329-333.

HAN J, ZHOU C, WU Y, et al. 2013. Self-assembling behavior of cellulose nanoparticles during freeze-drying: effect of suspension concentration, particle size, crystal structure, and surface charge [J]. Biomacromolecules, 14 (5): 1529-1540.

HASSAN M L, MATHEW A P, HASSAN E A, et al. 2012. Nanofibers from bagasse and rice straw: process optimization and properties [J]. Wood Science and Technology, 46 (1-3): 193-205.

HILL C A S, NORTON A J, NEWMAN G. 2010. The water vapor sorption properties of Sitka spruce determined using a dynamic vapor sorption apparatus [J]. Wood Science and Technology, 44 (3): 497-514.

HILL C A S, NORTON A, NEWMAN G. 2010. Analysis of the water vapor sorption behavior of Sitka spruce [Picea sitchensis (Bongard) Carr.] based on the parallel exponential kinetics model [J]. Holzforschung, 64 (4): 469-473.

HILL C A S, NORTON A, NEWMAN G. 2010. The water vapor sorption behavior of flax fibers-analysis using the parallel exponential kinetics model and determination of the activation energies of sorption [J]. Journal of Applied Polymer Science, 116 (4): 2166-2173.

HILL C A S, RAMSAY J, KEATING B, et al. 2012. The water vapor sorption properties of thermally modified and densified wood [J]. Journal of Materials Science, 47 (7): 3191-3197.

HOU L, FENG K, WU P, et al. 2014. Investigation of water diffusion process in ethyl cellulose-based films by attenuated total reflectance Fourier transform infrared spectroscopy and two-dimensional correlation analysis [J]. Cellulose, 21 (6): 4009-4017.

ISA A, MINAMINO J, MIZUNO H, et al. 2013. Increased water resistance of bamboo flour/polyethylene composites [J]. Journal of Wood Chemistry and Technology, 33 (3): 208-216.

KACHRIMANIS K, NOISTERNIG M F, GRIESSER U J, et al. 2006. Dynamic moisture sorption and desorption of standard and silicified microcrystalline cellulose [J]. European Journal of

Pharmaceutics and Biopharmaceutics, 64 (3): 307-315.

KHALIL H P S A, BHAT A H, YUSRA A F I. 2012. Green composites from sustainable cellulose nanofibrils: a review [J]. Carbohydrate Polymers, 87 (2): 963-979.

KHALIL H P S A, DAVOUDPOUR Y, SAURABH C K, et al. 2016. A review on nanocellulosic fibers as new material for sustainable packaging: process and applications [J]. Renewable and Sustainable Energy Reviews, 64: 823-836.

KITTLE J D, DU X, JIANG F, et al. 2011. Equilibrium water contents of cellulose films determined via solvent exchange and quartz crystal microbalance with dissipation monitoring [J]. Biomacromolecules, 12 (8): 2881-2887.

KOHLER R, DUCK R, AUSPERGER B, et al. 2003. A numeric model for the kinetics of water vapor sorption on cellulosic reinforcement fibers [J]. Composite Interfaces, 10 (2-3): 255-276.

LIANG C Y, MARCHESSAULT R H. 1959. Infrared spectra of crystalline polysaccharides .1. hydrogen bonds in native celluloses [J]. Journal of Polymer Science, 37 (132): 385-395.

LIU C, SHAO Z, WANG J, et al. 2016. Eco-friendly polyvinyl alcohol/cellulose nanofiber-Li^+ composite separator for high-performance lithium-ion batteries [J]. RSC Advances, 6 (100): 97912-97920.

LOUTHERBACK K, BIRARDA G, CHEN L, et al. 2016. Microfluidic approaches to synchrotron radiation-based Fourier transform infrared (SR-FTIR) spectral microscopy of living biosystems [J]. Protein and Peptide Letters, 23 (3): 273-282.

LUNDAHL M J, CUNHA A G, ROJO E, et al. 2016. Strength and water interactions of cellulose I filaments wet-spun from cellulose nanofibril hydrogels [J]. Scientific Reports, 6 (1): 30695.

MADAMBA P S, DRISCOLL R H, BUCKLE K A. 1996. The thin layer drying characteristics of garlic slices [J]. Journal of Food Engineering, 29 (1): 75-97.

MANHAS N, BALASUBRAMANIAN K, PRAJITH P, et al. 2015. PCL/PVA nanoencapsulated reinforcing fillers of steam exploded/autoclaved cellulose nanofibrils for tissue engineering applications [J]. RSC Advances, 5 (31): 23999-24008.

MAZZEO R, JOSEPH E. 2007. Attenuated total reflectance microspectroscopy mapping for the characterisation of bronze corrosion products [J]. European Journal of Mineralogy, 19 (3): 363-371.

MENON R S, MACKAY A L, HAILEY J, et al. 1987. An NMR determination of the physiological water distribution in wood during drying [J]. Journal of Applied Polymer Science, 33 (4): 1141-1155.

MURPHY D, DEPINHO M N. 1995. An ATR-FTIR study of water in cellulose-acetate membranes prepared by phase inversion [J]. Journal of Membrane Science, 106 (3): 245-257.

OGIWARA Y, KUBOTA H, HAYASHI S, et al. 1969. Studies of water adsorbed on cellulosic materials by a high resolution NMR spectrometer [J]. Journal of Applied Polymer Science, 13 (8):

1689-1695.

OKUBAYASHI S, GRIESSER U J, BECHTOLD T. 2005. Moisture sorption/desorption behavior of various manmade cellulosic fibers [J]. Journal of Applied Polymer Science, 97 (4): 1621-1625.

OLSSON A M, SALMEN L. 2004. The association of water to cellulose and hemicellulose in paper examined by FTIR spectroscopy [J]. Carbohydrate Research, 339 (4): 813-818.

PERESIN M S, HABIBI Y, VESTERINEN A, et al. 2010. Effect of moisture on electrospun nanofiber composites of poly(vinyl alcohol) and cellulose nanocrystals [J]. Biomacromolecules, 11 (9): 2471-2477.

POPESCU C, HILL C A S, CURLING S, et al. 2014. The water vapor sorption behavior of acetylated birch wood: how acetylation affects the sorption isotherm and accessible hydroxyl content [J]. Journal of Materials Science, 49 (5): 2362-2371.

RAUTKARI L, HILL C A S, CURLING S, et al. 2013. What is the role of the accessibility of wood hydroxyl groups in controlling moisture content [J]. Journal of Materials Science, 48 (18): 6352-6356.

ROSA M F, MEDEIROS E S, MALMONGE J A, et al. 2010. Cellulose nanowhiskers from coconut husk fibers:effect of preparation conditions on their thermal and morphological behavior [J]. Carbohydrate Polymers, 81 (1): 83-92.

RUAN R, LUN Y, ZHANG J, et al. 1996. Structure-function relationships of highly refined cellulose made from agricultural fibrous residues [J]. Applied Engineering in Agriculture, 12 (4): 465-468.

SHI X, ZHENG Y, WANG G, et al. 2014. PH and electro-responsive characteristics of bacterial cellulose nanofiber sodium alginate hybrid hydrogels for the dual controlled drug delivery [J]. RSC Advances, 4 (87): 47056-47065.

SMITH G. 1995. Dielectric analysis of water in microcrystalline cellulose [J]. Pharmacy and Pharmacology Communications, 1 (9): 419-422.

STELTE W, SANADI A R. 2009. Preparation and characterization of cellulose nanofibers from two commercial hardwood and softwood pulps [J]. Industrial and Engineering Chemistry Research, 48 (24): 11211-11219.

SUGIMOTO H, MIKI T, KANAYAMA K, et al. 2008. Dielectric relaxation of water adsorbed on cellulose [J]. Journal of Non-Crystalline Solids, 354 (27): 3220-3224.

SUN X, WU Q, REN S, et al. 2015. Comparison of highly transparent all-cellulose nanopaper prepared using sulfuric acid and TEMPO-mediated oxidation methods [J]. Cellulose, 22 (2): 1123-1133.

URAKI Y, MATSUMOTO C, HIRAI T, et al. 2010. Mechanical effect of acetic acid lignin adsorption on Honeycomb-Patterned cellulosic films [J]. Journal of Wood Chemistry and Technology, 30 (4): 348-359.

VOGT B D, SOLES C L, LEE H J, et al. 2005. Moisture absorption into ultrathin hydrophilic polymer films on different substrate surfaces [J]. Polymer, 46 (5): 1635-1642.

WAN Y Z, LUO H, HE F, et al. 2009. Mechanical, moisture absorption, and biodegradation behaviors of bacterial cellulose fibre-reinforced starch biocomposites [J]. Composites Science and Technology, 69 (7): 1212-1217.

WATANABE A, MORITA S, OZAKI Y. 2006. A study on water adsorption onto microcrystalline cellulose by near-infrared spectroscopy with two-dimensional correlation spectroscopy and principal component analysis [J]. Applied Spectroscopy, 60 (9): 1054-1061.

XIE Y, HILL C A S, JALALUDIN Z, et al. 2011. The dynamic water vapor sorption behavior of natural fibers and kinetic analysis using the parallel exponential kinetics model [J]. Journal of Materials Science, 46 (2): 479-489.

XIE Y, HILL C A S, JALALUDIN Z, et al. 2011. The water vapor sorption behavior of three celluloses: Analysis using parallel exponential kinetics and interpretation using the Kelvin-Voigt viscoelastic model [J]. Cellulose, 18 (3): 517-530.

XU Z, LI J, ZHOU H et al. 2016. Morphological and swelling behavior of cellulose nanofiber (CNF)/poly(vinyl alcohol) (PVA) hydrogels: poly(ethylene glycol) (PEG) as porogen [J]. RSC Advances, 6 (49): 43626-43633.

ZAIHAN J, HILL C A S, CURLING S, et al. 2009. Moisture adsorption isotherms of acacia mangium and endospermum malaccense using dynamic vapor sorption [J]. Journal of Tropical Forest Science, 21 (3): 277-285.

Chapter 5 Quantitative evaluation of moisture sorption in TEMPO oxidized cellulose using micro-FTIR spectroscopy

5.1 Introduction

Cellulose is a main component of plant cell walls and a widespread biopolymer (Hon 1994; Padalkar et al. 2010; Sampath et al. 2016), derived from trees, agricultural and sideline products, hemp fabrics, cotton, and animal fibers in marine life. In addition, cellulose is also considered to be a renewable resource derived from nature. It is currently the most studied and widely used component in the chemical composition of biomass. Cellulose has been constantly innovated, widely used in many traditional fields including composites (Bledzki and Gassan 1999; Whitney et al. 2000), netting (Kamel et al. 2008), upholstery (Joly et al. 1996), coatings (Kumar et al. 2016; Potjewijd et al. 1995), packing (Nicolas et al. 1999), paper (Basta and El-Saied 2009; Lavoine et al. 2014; Tarres et al. 2016), etc. With the continuous deepening of human environmental awareness, the promotion and application of cellulose and its derivatives will continue to become a hot spot. The novel derivation of cellulose material, nanocellulose film, has it's own desired physico-chemical properties, such as high mechanical strength (Mahfoudhi and Boufi 2017; Shimizu et al. 2013), low thermal expansion coefficient (Deepa et al. 2015; Puangsin et al. 2013), and environmental friendly (Osterberg et al. 2013). However, the nanocellulose film is inherently hygroscopic because it contains many hydrophilic groups, such as hydroxyl and carbonyl groups. Once the nanocellulose film is applied in the external environment (temperature and relative humidity may be changing), moisture adsorption or desorption may occur, and the varying moisture content results in the variation of these physico-chemical properties. For these properties are affected by the moisture content, characterizing water adsorption of nanocellulose film is crucial for the commercial utilization of this environmentally friendly film.

Water adsorption, the inherent property of cellulose material, has been studied from various perspectives (Alamri and Low 2012,2012; Bismarck et al. 2002; Mueller et al. 2009; Qian and Bogner 2011; Zhou et al. 2001). A large number of literatures have studied the water content and shrinkage rate, water relaxation time, equilibrium water content and adsorption isotherms of cellulose material. Among them, the sorption isotherm is one of the key perspectives (Pizzi et al. 1987; Torres et al. 2012; Wolf et al. 1984), which is mainly measured by gravimetric methods, especially dynamic vapor sorption (DVS) (Belbekhouche et al. 2011; Driemeier et al. 2012; Kulasinski et al. 2016; Passauer et al. 2012; Popescu et al. 2014). For example, using one analytical balance (with a precision of 0.1 mg), Wan *et al.* (Wan et al. 2009) studied bacterial cellulose fibre reinforced starch biocomposites whose dimension was 30 mm by 10 mm to obtain the moisture absorption curve. Belbekhouche *et al.* (Belbekhouche et al. 2011) placed microfibrillated cellulose and cellulose whisker films (10-40 mg) into a quartz sample pan of the DVS apparatus, and determined the water vapor sorption isotherms of these two sample films. In addition, Wu *et al.* (Wu et al. 2018) used the same method to measure the moisture adsorption isotherm of konjac glucomannan-ethyl cellulose film, and he only loaded 5 mg cellulose film sample on the tray of the DVS sample chamber. Further, using DVS apparatus, the sorption isotherms of other cellulose materials including regenerated cellulose film (Sun et al. 2010), nanofibrillated cellulose film (Ansari et al. 2014), microcrystalline cellulose (Xie et al. 2011), and cellulose nanocrystal and nanofiber films (Guo et al. 2017) were all determined. Although the results are promising, this technique also suffers from resolution-related issues regarding the size of the measured sample, the mass of the measured sample and the measurable moisture content range.

In order to go deeply into the researches on water adsorption in cellulose material, many spectroscopic methods such as near-infrared spectroscopy (Berthold et al. 1998; Delwiche et al. 1992; Inagaki et al. 2008; Shinzawa et al. 2013), Fourier transform infrared spectroscopy (Célino et al. 2014; Guo et al. 2017; Olsson and Salmen 2004) and Raman spectroscopy (Ding et al. 2016; Scherer et al. 1985) have been exploited to study this phenomenon at the molecular level. For example, Watanabe *et al.* (Watanabe et al. 2006) used near-infrared spectroscopy to study water adsorption on microcrystalline cellulose, and revealed the changes in the H-bonding interaction. Balcerzak and Mucha (Balcerzak and Mucha 2008) demonstrated that the FTIR spectroscopy has enough ability to qualitatively analyze water sorption of cellulose, and confirmed that the absorbance of hydroxyl group band was improved with the increase of adsorbed water. Agarwal and Kawai (Agarwal and Kawai 2005) examined the water sorption in the cellulose filter paper, determined a linear dependence relation between the water and Raman intensity of the O—H stretching envelope. Among these spectroscopic methods, FTIR spectroscopy plays an important role in component analysis and property prediction of organic compounds with hydrogen-containing groups. Cellulose, as a natural organic polymer material, contains a large number of hydrogen bonds and hydrogen-containing groups, which means that FTIR spectroscopy is a powerful analytical tool in studying the relationship between cellulose and moisture. At the same time, FTIR spectroscopy technology also has the following unique-advantages, such as the easily distinguishable characteristic peak (Abidi et al. 2014; Movasaghi et al. 2008), the high sensitivity of water (Meng et al. 2012), and the good accuracy of spectral analysis (Moreira and Santos 2004). Moreover, the development of micro-FTIR spectroscopy is convenient for extending this technique to micron-sized sample (Chen et al. 2015; Mastalerz and Bustin 1995). More recently, micro-FTIR spectroscopy equipped with an additional visible-light microscope that offers visual examination and selecting observation area has been introduced (Loutherback et al. 2016; Mazzeo and Joseph 2007). This technique yields high spatial resolution at the micron scale, and can provide *in situ* observation of water adsorption, which may feature multi-parameter measurement (Kohler et al. 2007), fast analysis (Li et al. 2007; Schulz and Kropp 1993), less sample consumption (Chonanant et al. 2011), and high resolution (Chen et al. 2013; Chen et al. 2015; Kijewski and Hofmann 1991; Yesiltas and Kebukawa 2016).

Considering that micro-FTIR spectroscopy has the capability to study water adsorption at the molecular level, in this book we proposed one strategy for qualitative and quantitative evaluation of water adsorption in nanogram quantities of nanocellulose film based on micro-FTIR spectroscopy and PLSR. Firstly, we collected micro-FTIR spectra of a typical nanocellulose film (i.e., TOCNF) over a wide range of RH levels which could be accurate controlled by a specially designed apparatus. Secondly, we qualitatively analyzed these spectra and corresponding difference spectra to confirm effective water sorption sites and identify the spectral ranges related to water adsorption. Thirdly, using the PLSR, we built and developed multivariate prediction models based on the identified the spectral ranges and the measured MCs. Finally, through comparative analyses between the estimated and measured MCs, we illustrated the feasibility of the proposed strategy.

5.2 Materials and methods

5.2.1 Materials

This book concentrated on the water adsorption of TOCNF. As the preparation for TOCNF was described clearly in previous literature (Isogai et al. 2011; Rodionova et al. 2012; Saito et al. 2007), the sample preparation was briefly described. Dried bleached wood pulp (W-50 grade from KC Flock, Nippon Paper Chemicals Co., Tokyo, Japan) acted as raw material. Bleached wood pulp (5g) was uniformly mixed with distilled water (150 mL). Then, NaBr (0.20g), 10 wt% NaClO solution (30g) and TEMPO (0.04g) were added successively. Meanwhile, the pH value of the mixed solution was adjusted to around 10.0 using 1 mol/L NaOH solution. The reaction time was set to 6 h. After the reaction completed, the ethanol (30 mL) was added. The obtained mixture was filtrated, and washed until the pH was equal to seven. 20 min later, the mixture was passed the homogenizer for 5 times. Then the concentration of this suspension was changed to 0.1 wt%. Subsequently, the suspension (10g) was dried in one culture dish with diameter of 60 mm at room temperature for four days to form TOCNF.

5.2.2 Experimental apparatus for micro-FTIR spectroscopy measurement

The experimental instrument is presented in Figure 5.1(a). The primary section of the instrument was a spectrometer (Nicolet IN 10^{TM}). This spectrometer was installed with an additional microscope, which was used for visual examination. In the spectral measurement, standard spectral acquisition was used for characterizing spectral development of TOCNF *vs.* RH. All micro-FTIR spectra were acquired from one randomly selected area (30μm by 30 μm, around 3 ng), and these spectra in the range of 720-4000 cm^{-1} were accumulated 32 times by scanning the grating at the spectral resolution of 4 cm^{-1}. Moreover, Figure 5.1(b) displays the schematic diagram of sample cell. Firstly, one TOCNF sample was placed on the bottom of this sample cell. Then, this cell was closed, and put on the automatic stage of the micro-FTIR spectrometer. Through the cell, Nitrogen gas with specific RH was circulated.

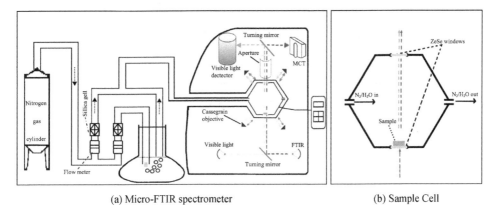

(a) Micro-FTIR spectrometer   (b) Sample Cell

Figure 5.1   Schematic diagram of experimental apparatus for micro-FTIR spectroscopy measurement

In this experiment, all spectral measurements were carried out during the RH range from 0 to 95% at the constant temperature of 25 ℃. Representative variations of RH and recorded micro-FTIR spectra *vs.* time were displayed in Figure 5.2. When target RH was set the next value (such as 5% RH and 10% RH), one delay time of 3-4 min would appear. During this delay time, actual RH got close to the target RH, and then remained stable throughout. After the 15 min, the spectra collected every 0.5 min stayed the same for 45 min (the height of the 3347 cm^{-1} peak closely associated with water adsorption was introduced to reveal the spectral change). Based on this, 60 min were chosen as balance time at each RH level.

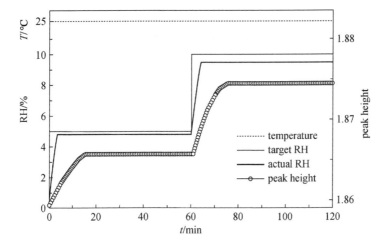

Figure 5.2   Representative variations of RH and micro-FTIR spectra of TOCNF against time during the measurement of micro-FTIR spectroscopy

5.2.3 Determination of moisture content using DVS apparatus

Moisture content (MC) of TOCNF was collected using the DVS apparatus (DVS AdvantagePlus). First, TOCNF sample was placed on a metal plate. Then this metal plate was hung in the microbalance which was fixed in the temperature and humidity auto-controlled box. The target RH was set to increase in 5% RH steps from 0% to 95%, and then reduce to 0% RH in the same step. Meanwhile, every target RH was maintained for enough time until sample mass altered less than 0.002% per minute during 10 min. In the sorption cycle, the temperature remained 25 °C.

The moisture content was computed based on the recorded sample mass and the following formula:

$$\mathrm{MC} = \frac{m - m_\mathrm{d}}{m_\mathrm{d}} \times 100\% \tag{5.1}$$

Where, MC was moisture content, m_d was dry mass, and m was the recorded sample mass.

Typical changes of MC and RH *vs.* time were displayed in Figure 5.3. Once the target RH was set a new value (e.g. 10% RH), the time delay of 4-10 min would appear. During this delay time, actual RH got close to target RH, and then remained stable throughout. Meanwhile, typical variation of MC with time followed an asymptotic curve, and the MC achieved the equilibrium value after infinite time of exposure at the target RH. Furthermore, it should be noticed that three replicates of these TOCNF specimens were adopted and each equilibrium moisture content collected as reference value at different RH levels was mean of three replicates.

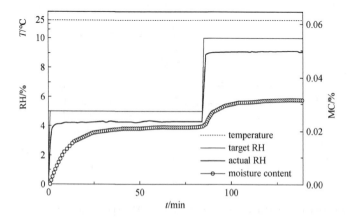

Figure 5.3 Representative variations of RH and MC of lignin against time during the measurement of moisture content using DVS apparatus

5.2.4 Data processing of micro-FTIR spectra

1. Acquisition of difference spectra

To quantitatively evaluating water sorption of TOCNF, FTIR difference spectrum technique was applied, by which difference spectrum was obtained by subtracting the spectrum measured at 0% RH.

2. Establishment of forecasting model based on micro-FTIR spectra

Multivariate quantitative forecasting model was established using PLSR in TQ Analyst 9 (Thermo Scientific Inc.). The pathlength type could be set to "constant". In components table, the only component, MC should be filled firstly, and maximum and minimum values of MC were set to "29%" and "0%" separately, which came from reference value collected by DVS apparatus. Then original micro-FTIR spectra of TOCNF measured at sixteen levels of RH were divided into two groups which were not modified by smoothing and baseline correction method. Spectral regions introduced to build PLSR model were adjusted according to the result of qualitatively characterizing water sorption of TOCNF, which would be described in detail later. Meanwhile, to avoid bias in these two groups, the original spectra collected at twelve levels of RH with five replicates (i.e., sixty spectra) were selected as calibration group. The remaining micro-FTIR spectra measured at four levels of RH with five replicates (i.e., twenty spectra) were applied into validation group. Using these spectral data in calibration group, calculations and cross-validations were carried out. After several iterations, the optical PLSR model was obtained, and the correlation of determination (R^2) and root mean square error of cross validation (RMSECV) were computed. Then using these spectral data in validation group, the model evaluation was implemented. Meanwhile, R^2 and root mean square error of prediction (RMSEP) were determined.

5.3 Results and discussion

5.3.1 Quantitative analysis of moisture adsorption in TOCNF

Figure 5.4 demonstrates TOCNF spectra measured in the RH region from 0 to 96%. The development of micro-FTIR spectra *vs.* RH was displayed. Meanwhile, the assignment of main bands in these recorded spectra associated with water adsorption to functional groups is summarized in Table 5.1.

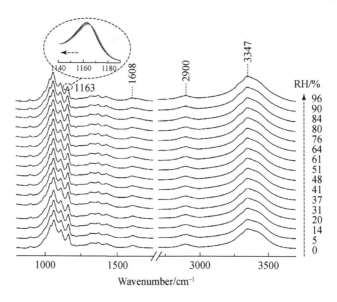

Figure 5.4 TOCNF spectra collected in the RH range from 0 to 96%

Table 5.1 Assignment of the main absorption bands in micro-FTIR spectra of TOCNF which were associated with water adsorption

Wavenumber /cm^{-1}	Assignment
3347	O—H stretching vibration
1608	C=O stretching vibration
1163	C—O—C asymmetric stretching vibration

With an increase in RH, the main band near 3347 cm^{-1} belonged to O—H stretching vibration was enlarged, indicating that water molecule was successively adsorbed in the OH group (Balcerzak and Mucha 2008; Velazquez et al. 2003). Meanwhile, the weaker band located next to 1608 cm^{-1} belonged to C=O stretching vibration, and it was also found to increase with RH. These changes suggested that some water molecule was adsorbed in the O—H and C=O groups. Furthermore, the 1163 cm^{-1} band belonged to C—O—C asymmetric stretching vibration was observed to shift toward lower wave numbers continuously with increasing RH. This shift was due to water adsorption, because the moisture adsorption stiffened cellulose chain, and then affected C—O—C asymmetric stretching vibration. Based on this, we could confirm that three spectral ranges of 3700-3000 cm^{-1}, 1700-1573 cm^{-1}, and 1180-1140 cm^{-1} were correlated with moisture adsorption.

In order to further extract precise spectral information about water adsorption, difference spectrum was employed here. Figure 5.5 presents difference spectra acquired by subtracting micro-FTIR spectrum collected at 0% RH. The broad band envelope contained many component bands (e.g. the 3343 cm^{-1} band, the 3285 cm^{-1} band, etc.) was observed to rise with an increase in RH which was due to water adsorption in the OH group of TOCNF. The same variation was also observed in other cellulose materials (Velazquez et al. 2003). Based on this fact, more accurate spectral range of this main band correlated with moisture adsorption was inferred as from 3700 cm^{-1} to 2880 cm^{-1}. Meanwhile, the 1650 cm^{-1} band assigned to adsorbed water displayed the same change tendency, and it further confirmed the second spectral range (1700-1573 cm^{-1}) resulted from moisture adsorption. Moreover, with an increase of RH, the intensity of the 1171 cm^{-1} band was shown to decrease, while the signature from the 1159 cm^{-1} band was found to rise. These negative-going and positive-going were due to the shift of the 1163 cm^{-1} in Figure 5.4. This further suggested the third range of 1180-1140 cm^{-1} was correlated with moisture adsorption.

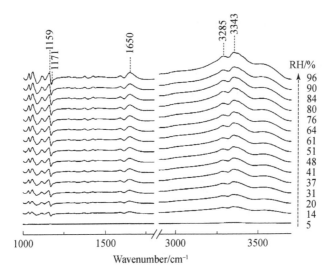

Figure 5.5   Difference spectra of TOCNF collected in the RH range from 5% to 96%

Moreover, peak height changes of these mentioned bands closely correlated with moisture sorption *vs*. RH are displayed in Figure 5.6. As shown in this figure, the growth trends of these three peak heights were different. Therefore, any of these peaks failed to determine the water sorption isotherm using the univariate analysis. This might be attributable to the fact that all sorption sites played their part in the water adsorption, rather than only one sorption site. Considering this reason, multivariate analysis model should be suitable for quantitatively evaluating water sorption in TOCNF.

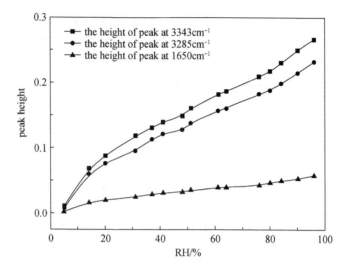

Figure 5.6　Peak height changes of micro-FTIR bands affected by water adsorption against RH

5.3.2　Quantitatively evaluating water adsorption of TOCNF

As described in the introduction, DVS has offered vast amounts of moisture sorption. Hence, this technique was used to collect MC of TOCNF as reference values. The recorded sorption isotherm was displayed in Figure 5.7, where the MC can be shown clearly. It was found that the MC curve against RH followed typical sigmoidal shape.

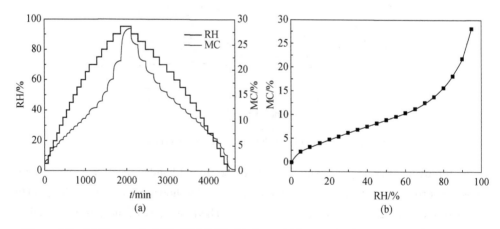

Figure 5.7　(a) Changes in MC of TOCNF with the variable over the time profile during isotherm runs; (b) MCs of TOCNF measured using DVS apparatus

In order to establish PLSR forecasting model for quantitatively evaluating water adsorption, the essential argument of spectral region should be made firstly. Three mentioned ranges of 3700-2800 cm^{-1}, 1700-1573 cm^{-1}, and 1180-1140 cm^{-1} were all introduced as first case. Meanwhile, the widened and narrowed spectral ranges were recommended separately as second and third case (Second case: 3700-2800 cm^{-1}, 1700-1573 cm^{-1}, 1180-1140 cm^{-1}, and 1556-1218 cm^{-1}; Third case: 3700-3000 cm^{-1}, 1700-1573 cm^{-1}, and 1180-1140 cm^{-1}). For comparison, three cases were all employed to build the PLSR quantitative model. For these three cases, the setting parameters such as selected spectral regions and the quality parameters including RESECV, RESEP, and R^2 were all listed in Table 5.2. The established model in the first case had maximum R^2, minimum RMSEP and RMSECV. It was confirmed that the first model owned high forecast precision. The best model took advantage of the complete spectral regions related to water adsorption (3700-2800 cm^{-1}, 1700-1573 cm^{-1}, and 1180-1140 cm^{-1}), while the widened and narrowed spectral regions (second case and third case) had a harmful effect on the performance of the established model.

Table 5.2 PLSR quality parameters for cross and test set validation for the proposed three cases

	First case	Second case	Third case
Spectral region/cm^{-1}	3700-2800/1700-1573/ 1180-1140	3700-2800/1700-1573/ 1180-1140/1556-1218	3700-3000/1700-1573/ 1180-1140
Cross validation			
RMSECV	0.253	0.280	0.303
Number of PLSR components	6	6	6
R^2 /%	99.95	99.94	99.92
External validation			
RMSEP	0.448	0.453	0.454
Number of PLSR components	6	6	6
R^2 /%	99.84	99.83	99.83

RMSECV, root mean square error of cross validation; R^2, coefficient of determination; RMSEP, root mean square error of prediction.

Taking advantage of the optical forecasting model, MCs of TOCNF at various RH levels were predicted. For comparison, the reference values measured by DVS apparatus were also displayed in Figure 5.8. As described in the literature, the moisture adsorption process of cellulosic material included three sections (Froix and Nelson 1975). In the first section, water molecule was absorbed on the hydroxyl and carboxyl groups. In the second section, the first hydration layer of these effective sorption sites was almost fully occupied, and the absorbed water was indirectly bond via another water molecule. In the third section, the absorbed water formed the five-molecule tetrahedral structure. In these three sections, the estimated MCs based on micro-FTIR spectra were quite close to the measured MCs (relative error was lower than 4%). Results indicated that this strategy for qualitative and quantitative evaluation of water adsorption in nanogram quantities of nanocellulose film was effective and efficient.

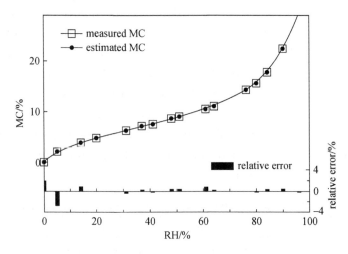

Figure 5.8 A comparison of sorption isotherm estimated by micro-FTIR forecasting model and that determined by DVS apparatus

5.4 Conclusions

Using micro-FTIR spectroscopy and PLSR, the strategy for qualitative and quantitative evaluation of water adsorption in nanogram quantities of TOCNF was demonstrated. TOCNF spectra were collected in the RH range from 0% to 96%. An analysis of these spectra and corresponding difference spectra confirmed effective sorption sites of TOCNF, including OH, and C= groups. Moreover, three micro-FTIR spectral regions correlated with moisture adsorption, consisting of 3700-2800 cm^{-1}, 1700-1573 cm^{-1}, and 1180-1140 cm^{-1} were identified. Based on these identified spectral regions and MCs measured by DVS apparatus, the micro-FTIR forecasting model was constructed using PLSR. Further, this micro-FTIR forecasting model was used to generate sorption isotherm of TOCNF which well matched that measured using DVS apparatus, and it confirmed the effectiveness of the strategy for qualitative and quantitative evaluation of water adsorption in nanogram quantities of TOCNF.

References

ABIDI N, CABRALES L, HAIGLER C H. 2014. Changes in the cell wall and cellulose content of developing cotton fibers investigated by FTIR spectroscopy [J]. Carbohydrate Polymers, 100: 9-16.

AGARWAL U P, KAWAI N. 2005. "Self-absorption" phenomenon in near-infrared Fourier transform Raman spectroscopy of cellulosic and lignocellulosic materials [J]. Applied Spectroscopy, 59 (3): 385-388.

ALAMRI H, LOW I M. 2012a. Effect of water absorption on the mechanical properties of n-SiC filled recycled cellulose fibre reinforced epoxy eco-nanocomposites [J]. Polymer Testing, 31 (6): 810-818.

ALAMRI H, LOW I M. 2012. Mechanical properties and water absorption behavior of recycled cellulose fibre reinforced epoxy composites [J]. Polymer Testing, 31 (5): 620-628.

ANSARI F, GALLAND S, JOHANSSON M, et al. 2014. Cellulose nanofiber network for moisture stable, strong and ductile biocomposites and increased epoxy curing rate [J]. Composites Part A-Applied Science and Manufacturing, 63: 35-44.

BALCERZAK J, MUCHA M. 2008. Study of adsorption and desorption heats of water in chitosan and its blends with hydroxypropylcellulose [J]. Molecular Crystals and Liquid Crystals, 484 (1): 465-472.

BASTA A H, EL-SAIED H. 2009. Performance of improved bacterial cellulose application in the production of functional paper [J]. Journal of Applied Microbiology, 107 (6): 2098-2107.

BELBEKHOUCHE S, BRAS J, SIQUEIRA G, et al. 2011. Water sorption behavior and gas barrier properties of cellulose whiskers and microfibrils films [J]. Carbohydrate Polymers, 83 (4): 1740-1748.

BERTHOLD J, OLSSON R, SALMEN L. 1998. Water sorption to hydroxyl and carboxylic acid groups in carboxymethylcellulose (CMC) studied with NIR-spectroscopy [J]. Cellulose, 5 (4): 281-298.

BISMARCK A, ARANBERRI-ASKARGORTA I, SPRINGER J, et al. 2002. Surface characterization of flax, hemp and cellulose fibers; Surface properties and the water uptake behavior [J]. Polymer Composites, 23 (5): 872-894.

BLEDZKI A K, GASSAN J. 1999. Composites reinforced with cellulose based fibers [J]. Progress in Polymer Science, 24 (2): 221-274.

CÉLINO A, GONCALVES O, JACQUEMIN F, et al. 2014. Qualitative and quantitative assessment of water sorption in natural fibers using ATR-FTIR spectroscopy [J]. Carbohydrate Polymers, 101: 163-170.

CHEN Y, CARO L D, MASTALERZ M, et al. 2013. Mapping the chemistry of resinite, funginite and associated vitrinite in coal with micro-FTIR [J]. Journal of Microscopy, 249 (1): 69-81.

CHEN Y, ZOU C, MASTALERZ M, et al. 2015. Applications of micro-Fourier transform infrared spectroscopy (FTIR) in the geological Sciences-A review [J]. International Journal of Molecular Sciences, 16 (12): 30223-30250.

CHONANANT C, JEARANAIKOON N, LEELAYUWAT C, et al. 2011. Characterisation of chondrogenic differentiation of human mesenchymal stem cells using synchrotron FTIR microspectroscopy [J]. Analyst, 136 (12): 2542-2551.

DEEPA B, ABRAHAM E, CORDEIRO N, et al. 2015. Utilization of various lignocellulosic biomass for the production of nanocellulose: a comparative study [J]. Cellulose, 22 (2): 1075-1090.

DELWICHE S R, NORRIS K H, PITT R E. 1992. Temperature sensitivity of near-infrared scattering transmittance spectra of water-adsorbed starch and cellulose [J]. Applied Spectroscopy, 46 (5): 782-789.

DING T, WANG C, PENG W. 2016. A theoretical study of moisture sorption behavior of heat-treated pine wood using Raman spectroscopic analysis [J]. Journal of Forestry Engineering, 1 (5): 15-19.

DRIEMEIER C, MENDES F M, OLIVEIRA M M. 2012. Dynamic vapor sorption and thermoporometry to probe water in celluloses [J]. Cellulose, 19 (4): 1051-1063.

FROIX M F, NELSON R. 1975. The interaction of water with cellulose from nuclear magnetic resonance relaxation times [J]. Macromolecules, 8 (6): 726-730.

GUO X, WU Y, XIE X. 2017. Water vapor sorption properties of cellulose nanocrystals and nanofibers using dynamic vapor sorption apparatus [J]. Scientific Reports, 7 (1): 14207.

GUO X, WU Y, YAN N. 2017. Characterizing spatial distribution of the adsorbed water in wood cell wall of Ginkgo biloba L. by μ-FTIR and confocal Raman spectroscopy [J]. Holzforschung, 71 (5): 415-423.

HON D. 1994. Cellulose - a random-walk along its historical path [J]. Cellulose, 1 (1): 1-25.

INAGAKI T, YONENOBU H, TSUCHIKAWA S. 2008. Near-infrared spectroscopic monitoring of the water adsorption/desorption process in modern and archaeological wood [J]. Applied Spectroscopy, 62 (8): 860-865.

ISOGAI A, SAITO T, FUKUZUMI H. 2011. TEMPO-oxidized cellulose nanofibers [J]. Nanoscale, 3 (1): 71-85.

JOLY C, KOFMAN M, GAUTHIER R. 1996. Polypropylene/cellulosic fiber composites: chemical treatment of the cellulose assuming compatibilization between the two materials [J]. Journal of Macromolecular Science-Pure and Applied Chemistry, 33 (12): 1981-1996.

KAMEL S, ALI N, JAHANGIR K, et al. 2008. Pharmaceutical significance of cellulose: a review [J]. Express Polymer Letters, 2 (11): 758-778.

KIJEWSKI H, HOFMANN M. 1991. FTIR-microspectrophotometry for high resolution and highly sensitive detection of the carboxyhemoglobin complex [J]. Beiträge Zur Gerichtlichen Medizin, 49: 137-141.

KOHLER A, BERTRAND D, MARTENS H, et al. 2007. Multivariate image analysis of a set of FTIR microspectroscopy images of aged bovine muscle tissue combining image and design information [J]. Analytical and Bioanalytical Chemistry, 389 (4): 1143-1153.

KULASINSKI K, SALMEN L, DEROME D, et al. 2016. Moisture adsorption of glucomannan and xylan hemicelluloses [J]. Cellulose, 23 (3): 1629-1637.

KUMAR V, ELFVING A, KOIVULA H, et al. 2016. Roll-to-roll processed cellulose nanofiber coatings [J]. Industrial and Engineering Chemistry Research, 55 (12): 3603-3613.

LAVOINE N, BRAS J, DESLOGES I. 2014. Mechanical and barrier properties of cardboard and 3D packaging coated with microfibrillated cellulose [J]. Journal of Applied Polymer Science, 131 (8): 40106.

LI Z, FREDERICKS P M, RINTOUL L, et al. 2007. Application of attenuated total reflectance micro-Fourier transform infrared (ATR-FTIR) spectroscopy to the study of coal macerals: examples from the Bowen Basin, Australia [J]. International Journal of Coal Geology, 70 (1-3): 87-94.

LOUTHERBACK K, BIRARDA G, CHEN L, et al. 2016. Microfluidic approaches to synchrotron radiation-based Fourier transform infrared (SR-FTIR) spectral microscopy of living biosystems [J]. Protein and Peptide Letters, 23 (3): 273-282.

MAHFOUDHI N, BOUFI S. 2017. Nanocellulose as a novel nanostructured adsorbent for environmental remediation: a review [J]. Cellulose, 24 (3): 1171-1197.

MASTALERZ M, BUSTIN R M. 1995. Application of reflectance micro-fourier transform-infrared spectrometry in studying coal macerals - comparison with other fourier-transform infrared techniques [J]. Fuel, 74 (4): 536-542.

MAZZEO R, JOSEPH E. 2007. Attenuated total reflectance microspectroscopy mapping for the characterisation of bronze corrosion products [J]. European Journal of Mineralogy, 19 (3): 363-371.

MENG X, SEDMAN J, van de VOORT F R. 2012. Improving the determination of moisture in edible oils by FTIR spectroscopy using acetonitrile extraction [J]. Food Chemistry, 135 (2): 722-729.

MOREIRA J L, SANTOS L. 2004. Spectroscopic interferences in fourier transform infrared wine analysis [J]. Analytica Chimica Acta, 513 (1): 263-268.

MOVASAGHI Z, REHMAN S, REHMAN I U. 2008. Fourier transform infrared (FTIR) spectroscopy of biological tissues [J]. Applied Spectroscopy Reviews, 43 (2): 134-179.

MUELLER C M O, LAURINDO J B, YAMASHITA F. 2009. Effect of cellulose fibers addition on the mechanical properties and water vapor barrier of starch-based films [J]. Food Hydrocolloids, 23 (5): 1328-1333.

NICOLAS V, CHAMBIN O, ANDRES C, et al. 1999. Preformulation: effect of moisture content on microcrystalline cellulose (Avicel PH-302) and its consequences on packing performances [J]. Drug Development and Industrial Pharmacy, 25 (10): 1137-1142.

OLSSON A M, SALMEN L. 2004. The association of water to cellulose and hemicellulose in paper examined by FTIR spectroscopy [J]. Carbohydrate Research, 339 (4): 813-818.

OSTERBERG M, VARTIAINEN J, LUCENIUS J, et al. 2013. A fast method to produce strong NFC films as a platform for barrier and functional materials [J]. Acs Applied Materials and Interfaces, 5 (11): 4640-4647.

PADALKAR S, CAPADONA J R, ROWAN S J, et al. 2010. Natural biopolymers: novel templates for the synthesis of nanostructures [J]. Langmuir, 26 (11): 8497-8502.

PASSAUER L, STRUCH M, SCHULDT S et al. 2012. Dynamic moisture sorption characteristics of xerogels from water-swellable oligo(oxyethylene) lignin derivatives [J]. Acs Applied Materials and Interfaces, 4 (11): 5852-5862.

PIZZI A, BARISKA M, EATON N J. 1987. Theoretical water sorption energies by conformational-analysis .2. Amorphous cellulose and the sorption isotherm [J]. Wood Science and Technology, 21 (4): 317-327.

POPESCU C, HILL C A S, CURLING S, et al. 2014. The water vapor sorption behavior of acetylated birch wood: how acetylation affects the sorption isotherm and accessible hydroxyl content [J]. Journal of Materials Science, 49 (5): 2362-2371.

POTJEWIJD R, NISPEROS M O, BURNS J K, et al. 1995. Cellulose-based coatings as carriers for candida guillermondii and debaryomyces sp in reducing decay of oranges [J]. Hortscience, 30 (7): 1417-1421.

PUANGSIN B, YANG Q, SAITO T, et al. 2013. Comparative characterization of TEMPO-oxidized cellulose nanofibril films prepared from non-wood resources [J]. International Journal of Biological Macromolecules, 59: 208-213.

QIAN K K, BOGNER R H. 2011. Differential heat of adsorption of water vapor on silicified microcrystalline cellulose (SMCC): an investigation using isothermal microcalorimetry [J]. Pharmaceutical Development and Technology, 16 (6): 616-626.

RODIONOVA G, ERIKSEN O, GREGERSEN O. 2012. TEMPO-oxidized cellulose nanofiber films: effect of surface morphology on water resistance [J]. Cellulose, 19 (4): 1115-1123.

SAITO T, KIMURA S, NISHIYAMA Y, et al. 2007. Cellulose nanofibers prepared by TEMPO-mediated oxidation of native cellulose [J]. Biomacromolecules, 8 (8): 2485-2491.

SAMPATH U G T M, CHING Y C, CHUAH C H, et al. 2016. Fabrication of porous materials from natural/synthetic biopolymers and their composites [J]. Materials, 9 (12): 991.

SCHERER J R, BAILEY G F, KINT S, et al. 1985. Water in polymer membranes. 4. Raman scattering from cellulose acetate films [J]. Journal of Physical Chemistry, 89 (2): 312-319.

SCHULZ H, KROPP B. 1993. Micro spectroscopy FTIR reflectance examination of paint binders on ground chalk [J]. Fresenius' Journal of Analytical Chemistry, 346 (1): 114-122.

SHIMIZU M, FUKUZUMI H, SAITO T, et al. 2013. Preparation and characterization of TEMPO-oxidized cellulose nanofibrils with ammonium carboxylate groups [J]. International Journal of Biological Macromolecules, 59: 99-104.

SHINZAWA H, AWA K, NODA I, et al. 2013. Multiple-perturbation two-dimensional near-infrared correlation study of time-dependent water absorption behavior of cellulose affected by pressure [J]. Applied Spectroscopy, 67 (2): 163-170.

SUN S, MITCHELL J R, MACNAUGHTAN W, et al. 2010. Comparison of the mechanical properties of cellulose and starch films [J]. Biomacromolecules, 11 (1): 126-132.

TARRES Q, SAGUER E, PELACH M A, et al. 2016. The feasibility of incorporating cellulose micro/nanofibers in papermaking processes: The relevance of enzymatic hydrolysis [J]. Cellulose, 23 (2): 1433-1445.

TORRES M D, MOREIRA R, CHENLO F, et al. 2012. Water adsorption isotherms of carboxymethyl cellulose, guar, locust bean, tragacanth and xanthan gums [J]. Carbohydrate Polymers, 89 (2): 592-598.

VELAZQUEZ G, HERRERA-GOMEZ A, MARTIN-POLO M O. 2003. Identification of bound water through infrared spectroscopy in methylcellulose [J]. Journal of Food Engineering, 59 (1): 79-84.

WAN Y Z, LUO H, HE F, et al. 2009. Mechanical, moisture absorption, and biodegradation behaviors of bacterial cellulose fibre-reinforced starch biocomposites [J]. Composites Science and Technology, 69 (7-8): 1212-1217.

WATANABE A, MORITA S, OZAKI Y. 2006. A study on water adsorption onto microcrystalline cellulose by near-infrared spectroscopy with two-dimensional correlation spectroscopy and principal component analysis [J]. Applied Spectroscopy, 60 (9): 1054-1061.

WHITNEY S, GIDLEY M J, MCQUEEN-MASON S J. 2000. Probing expansin action using cellulose/hemicellulose composites [J]. Plant Journal, 22 (4): 327-334.

WOLF W, SPIESS W E L, JUNG G, et al. 1984. The water-vapor sorption isotherms of microcrystalline cellulose (MCC) and of purified potato starch. results of a collaborative study [J]. Journal of Food Engineering, 3 (1): 51-73.

WU K, ZHU Q, QIAN H, et al. 2018. Controllable hydrophilicity-hydrophobicity and related properties of konjac glucomannan and ethyl cellulose composite films [J]. Food Hydrocolloids, 79: 301-309.

XIE Y, HILL C A S, JALALUDIN Z, et al. 2011. The water vapor sorption behavior of three celluloses: Analysis using parallel exponential kinetics and interpretation using the Kelvin-Voigt viscoelastic model [J]. Cellulose, 18 (3): 517-530.

YESILTAS M, KEBUKAWA Y. 2016. Associations of organic matter with minerals in Tagish Lake meteorite via high spatial resolution synchrotron-based FTIR microspectroscopy [J]. Meteoritics and Planetary Science, 51 (3): 584-595.

ZHOU S, TASHIRO K, HONGO T, et al. 2001. Influence of water on structure and. mechanical properties of regenerated cellulose studied by an organized combination of infrared spectra, X-ray diffraction, and dynamic viscoelastic data measured as functions of temperature and humidity [J]. Macromolecules, 34 (5): 1274-1280.

Chapter 6 Molecular association of water with wood cell walls during moisture desorption process examined by micro- FTIR spectroscopy

6.1 Introduction

Wood is a natural polymer composite material, due to its renewable, recyclable, high strength to weight ratio, beautiful texture, and good acoustic, electrical and thermal properties, it can be used as an important engineering and construction material. At the same time, wood is a porous and hygroscopic biological material (Esteban et al. 2006; Wang and Wang 1999; Zhang et al. 2016). The hygroscopicity of wood is mainly due to the hydrophilic groups (mainly hydroxyl and carbonyl groups) in cellulose, lignin and hemicellulose. As wood is placed in high humidity condition, the moisture enters the internal cell wall of the wood in the form of gas, and forms a hydrogen bond with the free hydroxyl groups on the cell wall, thereby allowing the wood to exchange moisture with the atmospheric environment (hygroscopicity and desorption) to achieve internal moisture balance (Eligon et al. 1992; Guevara and Moslemi 1983; Ma et al. 2005; Merakeb et al. 2009). Besides, the varying moisture content of wood could result in the variation of dimensional stability (Srinivas and Pandey 2012), mechanical (Borrega and Karenlampi 2008; Gilani et al. 2014; Hosseinaei et al. 2012; Moliński and Raczkowski 1988; Obataya et al. 1998; Tamrakar and Lopez-Anido 2011), elastic (Hogan Jr and Niklas 2004; Maeda and Fukada 1987; Ozyhar et al. 2013), and electrical properties (Khan et al. 1991), etc. Since these important physical properties are relied on moisture content, a large number of literature have studied the theory of wood moisture absorption. Generally, the hygroscopic theory of wood mainly has the following aspects: First, the saturation theory of hydroxyl cohesion inside the micelles in the cellulose crystalline area. The second is that the free hydroxyl polar molecules in the amorphous region of cellulose are easy to combine with water to form a hydrogen bond theory. Third, the moisture absorption of the main components inside the wood is hemicellulose with the strongest hygroscopicity, followed by lignin, and the smallest cellulose.

Based on the previous theory of wood moisture absorption, many researchers have studied the moisture absorption properties of wood by using a number of experimental approaches (Engelund et al. 2010; Jalaludin et al. 2010; Joerdens et al. 2010; Nakano 2006; Taniguchi et al. 1978; Zhang 2011), such as dynamic vapor sorption (DVS) (Glass et al. 2017; Hill et al. 2012; Popescu et al. 2014; Rautkari et al. 2013; Xie et al. 2011), nuclear magnetic resonance (NMR) spectroscopy (Araujo et al. 1992; Cox et al. 2010; Hall and Rajanayagam 1986; Menon et al. 1989; Riggin et al. 1979), near-infrared (NIR) spectroscopy (Fujimoto et al. 2012; Koumbi-Mounanga et al. 2015; Leblon et al. 2013; Lestander and Geladi 2003; Mora et al. 2011; Nascimbem et al. 2013; Thygesen and Lundqvist 2000), and Fourier transform infrared (FTIR) spectroscopy (Akerholm and Salmen 2004; Guo et al. 2017; Olsson and Salmen 2004), etc. Among these, DVS is a quantitative analysis approach, which can rapidly record quality changes of sample in real time, and it has been used to analyze water vapor adsorption behavior and sorption isotherms of multifarious wood species including Scots pine (Altgen et al. 2016; Xie et al. 2010), Sitka spruce (Hill et al. 2010), and Malaysian hardwoods (Zaihan et al. 2009). Therefore, the dynamic vapor adsorption (DVS) device can not only determine the adsorption kinetics of the sample, but also measure the number of accessible hydroxyl groups by increasing the sample mass.

In addition to the dynamic vapor adsorption (DVS) device, near-infrared (NIR) spectroscopy is also a potential method to study the hygroscopic properties of wood. NIR spectroscopy is one analysis approach which can be used to investigate moisture sorption of wood using the spectral fingerprints of molecular overtone and combination vibrations of C—H and O—H function groups. Moreover, NMR technique is base on a physical phenomenon in which nuclei can absorb or send out electromagnetic radiation (Moudgil et al. 1985), by which the bound water and lumen water in wood can be distinguished. Further, FTIR spectroscopy is an effective molecular vibrational spectroscopic approach for functional group analysis (Somarathna et al. 2016), which has been successfully applied in qualitatively and quantitatively analyzing moisture adsorption in cellulosic materials (Dias et al. 1998; Fengel 2009; Hofstetter et al. 2006; Murphy and Pinho 1995). Among these approaches, FTIR spectroscopy provides some obvious advantages, i.e. the easily distinguishable characteristic peak (Abidi et al. 2014; Capron et al. 2007; Song et al. 2016), the high sensitivity of water (Ping et al. 2001;

Tian et al. 2016; Zhang et al. 2017), and the good accuracy of spectral analysis (Moreira and Santos 2004). Moreover, FTIR spectroscopy technology has the characteristics of greening and digitization as a detection method with rapid progress in recent years, and it's visible light band is easy to obtain, which contains a lot of scientific research information. More recently, micro-FTIR spectroscopy has been improved to own higher sensitivity in investigating molecular structure alteration. Based on this, micro-FTIR spectroscopy has largest potential to characterize moisture sorption in wood.

It is generally accepted that FTIR spectroscopy has offered lots of valuable results about moisture sorption. For instance, based on the FTIR spectra recorded at different RH levels, Chang *et al.* (2002) demonstrated that moisture content of raw wood was higher than that of acylated wood. Poletto *et al.* (Poletto et al. 2012) also used the same technique to research wood species, and found that the hydrogen bond energy at 3567 cm^{-1} could indicate the amount of absorbed water. In addition, Kondo (1997) showed that FTIR absorption bands were sensitive enough to determine whether the OH group bonded with water. Balcerzak *et al.* (Balcerzak and Mucha 2008) found that the absorbance of FTIR band of hydroxyl group was improved with the increase of adsorbed water. Moreover, Olsson *et al.* (2004) examined the adsorption characteristics of some wood polymers by FTIR spectroscopy, confirmed that the potential adsorption sites were the hydroxyl groups and the carboxyl groups, and indicated that a cluster type of water adhered to all types of potential adsorption sites. Célino *et al.* (2014) demonstrated that FTIR spectra collected at different RH levels could be used to develop multivariate model for predicting moisture content. Furthermore, Xu *et al.* (2013) showed that the spectral information of FTIR spectra could be used to qualitatively and quantitatively analyze moisture sorption mechanism. As mentioned, wood has complex chemical constituents that contain several potential water sorption sites (Peralta 1995; Xie et al. 2011; Xie et al. 2010). Due to various hydrogen-bond formations based on different sorption sites and molecular structure of desorbed water differ from general water, molecular structure change of water in wood during moisture sorption process is highly complex, which needs to be further investigated.

As previously described, micro-FTIR spectroscopy is considered to be an effective technique for qualitatively and quantitatively analyzing water sorption. Therefore, in this study we applied this vibrational spectroscopic technique to investigate the change of water molecular structure during moisture desorption process. Firstly, we collected wood spectra in the RH range from high level to low level.

Secondly, we analyzed these measured spectra to determine effective water sorption sites and identified the spectral region closely associated with moisture desorption. Finally, we applied the component peak analysis into the identified spectral region to demonstrate molecular structure change of desorbed water during the moisture desorption process.

6.2 Materials and methods

6.2.1 Materials

Wood specimens (dimensions 100 mm× 30 mm× 20 mm in length, width, and thickness) were cut from straight stem of Ginkgo biloba L. (Ginkgoaceae) wood. From these wood specimens, transverse sections were prepared without embedding and any chemical treatment. These sections were cut using a manual rotary microtome (Leica RM2135), and then caught between two slides. Furthermore, drying at (102 ± 3) ℃ was performed at least three hours before micro-FTIR spectral measurement.

6.2.2 Experimental instrument for micro-FTIR spectral measurement

The experimental instrument is presented in Figure 6.1. The primary section was a spectrometer (Nicolet IN 10^{TM}), which was applied for recording micro-FTIR spectral development of wood *vs.* RH during the RH range from 95% to 0% at the constant temperature of 25 ℃. This spectrometer was fitted with an additional microscope. In the spectral measurement, standard spectral acquisition was adopted, all micro-FTIR spectra were acquired from one randomly selected area (100 μm by 100 μm), and these spectra in the range of 720-4000 cm^{-1} were accumulated 32 times by scanning the grating at the spectral resolution of 4 cm^{-1}. Figure 6.1(b) also displays the schematic diagram of sample cell. Firstly, the wood transverse section was placed on the bottom of this sample cell. Then, this cell was closed, and put on the automatic stage of micro- FTIR spectrometer. The RH was changed using saturated steam-air mixture which was accurately controlled by flow meter. More, the RH could be tested using humidity and temperature meter and real-time displayed on the computer.

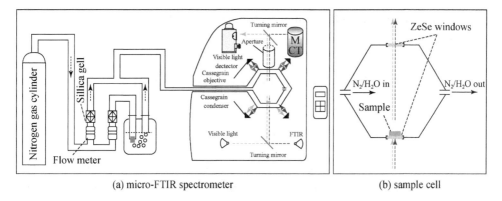

Figure 6.1　Experimental apparatus for spectral measurement

Prior to recording micro-FTIR spectral development of wood vs. RH, kinetic spectroscopy test was carried out to determine balance time. Once the target RH was set a new value, and a delay of about five minutes would occur. During this delay, actual RH became close to the target RH, and then remained almost unchanged afterwards. Meanwhile, after fifteen minutes, the spectra recorded every one minute were kept constant. Based on the above results, sixty minutes were set to balance time for micro-FTIR spectral measurement during each RH level.

6.2.3　Micro-FTIR spectral data processing

1. Achievement of difference spectra at various RH levels

To further qualitatively analyze moisture desorption in wood, FTIR difference spectrum technique was applied, using which difference spectra were obtained via the original micro-FTIR spectra collected at different RH levels subtracting the spectrum measured at 0% RH.

2. Component peak analysis used for difference spectra

Furthermore, component peak analysis was applied in distinguishing between different types of bond water. In the process of implementation, the identified spectral region was fitted using the Pseudo-Voigt function in Origin 7.0 software. This function was linear combination of Gaussian function and Lorentzian function, as shown.

$$y = y_0 + A\left[m_u \cdot \frac{2}{\pi} \cdot \frac{w}{4(x-x_c)^2 + w^2} + (1-m_u) \cdot \frac{\sqrt{4\ln 2}}{\sqrt{\pi}w} \cdot e^{-4\ln 2/w^2 \cdot (x-x_c)^2} \right] \quad (6.1)$$

In this equation, x was the independent variable, y was dependent variable, y_0 was baseline, A was peak area of the component peak, x_c was peak position of the component peak, w was full width at half-maximum of the component peak, and m_u was shape factor.

6.3 Results and discussion

6.3.1 Effective water sorption sites of wood

Figure 6.2 demonstrates micro-FTIR spectra measured in the RH region of 0-96%. As shown in this figure, the development of micro-FTIR spectra with RH was also shown clearly. Meanwhile, the assignment of main bands in these recorded spectra associated with moisture desorption to functional groups is listed in Table 6.1.

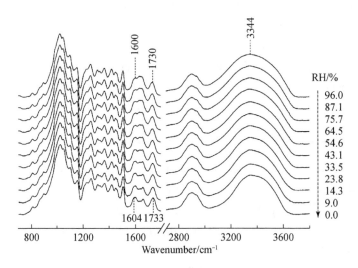

Figure 6.2 Micro-FTIR spectra collected over a wide range of RH from 0 to 96.0%

Table 6.1 Assignment of the main bands in micro-FTIR spectra of wood (collected at 96.0% RH) closely associated with moisture desorption

Wavenumber /cm^{-1}	Assignment
3700-3000	O—H stretching vibration
1730	C=O stretching vibration
1600	aromatic skeletal plus C=O stretching vibration

With a decrease in RH, one main band near 3344 cm^{-1} belonged to O—H stretching vibration was reduced, indicating that water was successively desorbed from the OH group. In the spectrum collected at 96.0% RH, two bands situated at 1730 and 1600 cm^{-1} belonged to C=O stretching vibration and aromatic skeletal plus C=O stretching vibration. For comparison, the spectrum of wood measured at 0 RH was also illustrated, in which these three bands appeared at 1733, and 1604 cm^{-1}. As shown, these two bands exhibited blue shifts, which inferred that some water was desorbed from the C=O group.

In order to further extract precise spectral information about moisture desorption, difference spectrum was employed. Figure 6.3 presents difference spectra collected during the RH decreased from 96.0% to 9.0%. As shown in this figure, the broad band situated between 3700 cm^{-1} and 2900 cm^{-1} was reduced *vs.* RH which was due to water desorption from the OH groups of wood. Based on this, it was inferred that more accurate spectral range of this main band correlated with moisture desorption was from 3700 cm^{-1} to 2900 cm^{-1}. Meanwhile, the 1754 cm^{-1} peak belonged to free C=O group increased, while the 1724 cm^{-1} peak attributed to hydrogen bonded C=O group had reverse trend. Accordingly, carbonyl group in wood was closely associated with moisture desorption.

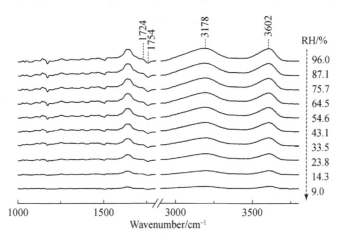

Figure 6.3　Difference spectra of wood collected during the RH decreased from 96.0% to 9.0%

To summarize, the results showed that water sorption sites included OH, and C=O groups, and it provided foundation for demonstrating molecular structures of desorbed water during moisture desorption process.

6.3.2 Molecular structure change of water during moisture desorption process

Previous studies confirmed that the O—H stretching envelope in FTIR spectroscopy is broad, unresolved and generated from various type of water. Therefore, component peak analysis was operated in this envelope situated between 3700 cm^{-1} and 2900 cm^{-1} to distinguish different types of water, as described in the materials and methods section. Here, three components were adopted. Among these components, two component peaks such as strongly and weakly bound waters were confirmed by Olsson *et al.*, whose parameters were introduced as initial values. More, the third component was lead to acquire more satisfactory fitting results, which was assigned to moderately bound water. The presence of moderately bond water was confirmed, therefore, the utilization of this component was reasonable. From component peak analysis, three component peaks were identified. The difference spectra collected at various RH levels during moisture desorption process as dotted lines and the corresponding fitted bands as solid lines are shown in Figure 6.4, in which three component peaks as dashed lines, RH and the coefficient of determination (R^2) are also demonstrated. In our fitting, the three component peaks were situated at 3178 cm^{-1}, 3514 cm^{-1} and 3602 cm^{-1} separately, and all of them had the same positions with those documented in the literature. The frequency of FTIR spectral peak is affected by hydrogen bonding, and some peak can display a red shift along with the enhancement of hydrogen-bonded association. Therefore, the component peak in the high frequency region (3178 wavenumber) was attributed to the strongly bond water, the component peak in moderate frequency region (3514 wavenumber) belonged to the moderately bond water, and the component peak in low frequency region (3602 wavenumber) belonged to the weakly bond water.

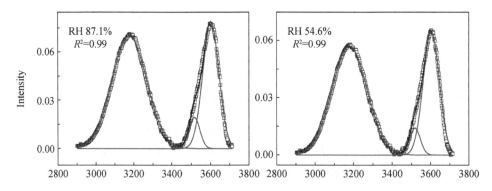

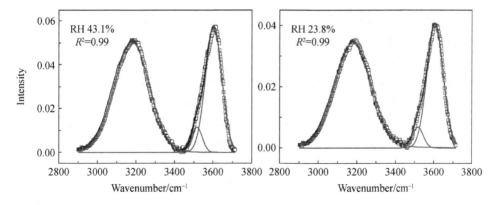

Figure 6.4 Component peak analysis applied in difference spectra

For demonstrating how three types of water altered during moisture desorption process, three component peak areas against RH are all exhibited. It was expected that the integrated intensities of these component peaks were reduced with the decrease of RH, for the water was desorbed during moisture desorption process. Meanwhile, there were obviously different reducing trends for these component peaks. According to these reducing trends, the moisture desorption process could be distinguished into three parts (Part One, Part Two and Part Three), as illustrated in Figure 6.5.

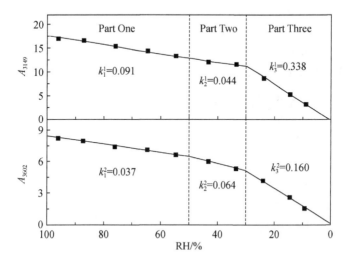

Figure 6.5 The amount of three types of water changed with RH during moisture desorption process, and k was the slope

In Part One, RH varied from 96.0 to 50.0%. The areas of two component peaks (A_{3178} and A_{3602}) both decreased, and the reducing rate of A_{3178} was higher than that of A_{3602}. Meanwhile, the area of the component peak at 3514 cm^{-1} remained almost unchanged. These changes indicated that the desorbed strongly bond water was more than the desorbed weakly bond water in this part. Previous literatures found that under high RH condition, water molecules are connected together with strong hydrogen bonds, and form five-molecule tetrahedral structure. Based on that the desorption happened at higher RH level, and strongly bond water preferred to desorbed in this part, the desorbed water mainly owned the five-molecule tetrahedral structure, as illustrated in Figure 6.6 (a).

In Part Two, the RH was in the range from 50.0% to 30.0%. The areas of two component peaks (A_{3178} and A_{3514}) were almost constant, while the area of the component peak (A_{3602}) decreased. Based on this, it inferred that main desorbed water in this part was weakly bond water. Therefore, in this part, the new desorbed water were those indirectly bond to the wood cell wall through the existing water, and their structures should be (WATER)···HOH, as illustrated in Figure 6.6 (b).

In Part Three, the RH was lower than 30.0%. All integrated intensities of three component peaks decreased, and the reducing rate of A_{3178} was highest. These changes confirmed strongly bond water was mostly desorbed in this part. As the moisture desorption happened at such lower RH, the desorbed water was that those bounded straightly at effective water sorption sites, which were proved to be carbonyl and hydroxyl groups. Therefore, in this part, most of desorbed water was $C=O \cdots (HOH)$ and $OH \cdots (OH_2)$ [Figure 6.6 (c)].

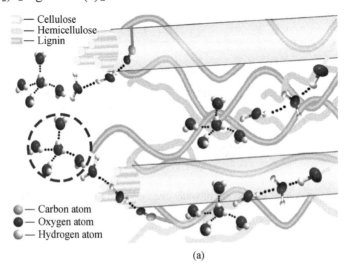

(a)

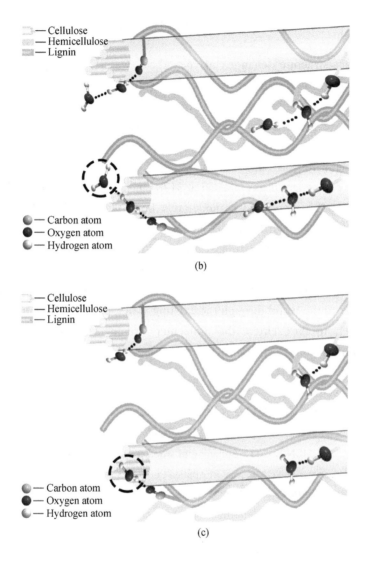

Figure 6.6 (a) The change of water molecular structure in Part One of moisture desorption process. When RH was from 96.0% to 50.0%, most of desorbed water was water cluster (as indicated in the black circle); (b) The change of water molecular structure in Part Two of moisture desorption process. As RH was in the range from 50.0% to 30.0%, most of desorbed water was weakly bond water (as indicated in the black circle); (c) The change of water molecular structure in Part Three of moisture desorption process. When RH was lower than 30.0%, most of desorbed water was strongly bond water (as indicated in the black circle)

6.4 Conclusions

Molecular structure change of desorbed water was investigated using micro-FTIR spectroscopy during moisture desorption process of wood. An analysis of these spectra and corresponding difference spectra indicated carbonyl and hydroxyl groups were the effective water sorption sites and identified spectral region between 3700 cm^{-1} and 2900 cm^{-1} was closely associated with moisture desorption. Moreover, component peak analysis was carried over into this identified spectral range, in which three component peaks situated at 3178 cm^{-1}, 3514 cm^{-1} and 3602 cm^{-1} were attributed to strongly bond water, moderately bond water and weakly bond water. There were obviously different reducing trends for these component peaks. According to these reducing trends, the moisture desorption process could be separated into three parts. In Part One, the desorbed water mainly owned the five-molecule tetrahedral structure; In Part Two, the desorbed water was those indirectly bond to the wood through the existing water, and their structures should be (water)···HOH; In Part Three, the molecular structures of desorbed water were $C = O···(HOH)$ and $OH···(OH_2)$.

References

ABIDI N, CABRALES L, HAIGLER C H. 2014. Changes in the cell wall and cellulose content of developing cotton fibers investigated by FTIR spectroscopy [J]. Carbohydrate Polymers, 100: 9-16.

AKERHOLM M, SALMEN L. 2004. Softening of wood polymers induced by moisture studied by dynamic FTIR spectroscopy [J]. Journal of Applied Polymer Science, 94 (5): 2032-2040.

ALTGEN M, HOFMANN T, MILITZ H. 2016. Wood moisture content during the thermal modification process affects the improvement in hygroscopicity of Scots pine sapwood [J]. Wood Science and Technology, 50 (6): 1181-1195.

ARAUJO C D, MACKAY A L, HAILEY J R T, et al. 1992. Proton magnetic resonance techniques for characterization of water in wood: application to white spruce [J]. Wood Science and Technology, 26 (2): 101-113.

BALCERZAK J, MUCHA M. 2008. Study of adsorption and desorption heats of water in chitosan and its blends with hydroxypropylcellulose [J]. Molecular Crystals and Liquid Crystals, 484 (1): 465-472.

BORREGA M, KARENLAMPI P P. 2008. Mechanical behavior of heat-treated spruce (Picea abies) wood at constant moisture content and ambient humidity [J]. Holz als Roh und Werkstoff, 66 (1): 63-69.

CAPRON I, ROBERT P, COLONNA P, et al. 2007. Starch in rubbery and glassy states by FTIR spectroscopy [J]. Carbohydrate Polymers, 68 (2): 249-259.

CÉLINO A, GONCALVES O, JACQUEMIN F, et al. 2014. Qualitative and quantitative assessment of water sorption in natural fibers using ATR-FTIR spectroscopy [J]. Carbohydrate Polymers, 101: 163-170.

CHANG H T, CHANG S T. 2002. Moisture excluding efficiency and dimensional stability of wood improved by acylation [J]. Bioresource Technology, 85 (2): 201-204.

COX J, MCDONALD P J, GARDINER B A. 2010. A study of water exchange in wood by means of 2D NMR relaxation correlation and exchange [J]. Holzforschung, 64 (2): 259-266.

DIAS C R, ROSA M J, de PINHO M N. 1998. Structure of water in asymmetric cellulose ester membranes — and ATR-FTIR study [J]. Journal of Membrane Science, 138 (2): 259-267.

ELIGON A M, ACHONG A, SAUNDERS R. 1992. Moisture adsorption and desorption properties of some tropical woods [J]. Journal of Materials Science, 27 (13): 3442-3456.

ENGELUND E T, THYGESEN L G, HOFFMEYER P. 2010. Water sorption in wood and modified wood at high values of relative humidity. Part 2: Appendix. theoretical assessment of the amount of capillary water in wood microvoids [J]. Holzforschung, 64 (3): 325-330.

ESTEBAN L G, FERNANDEZ F G, CASASUS A G, et al. 2006. Comparison of the hygroscopic behavior of 205-year-old and recently cut juvenile wood from Pinus sylvestris L. [J]. Annals of Forest Science, 63 (3): 309-317.

FENGEL D. 2009. Influence of water on the OH valency range in deconvoluted FTIR spectra of cellulose [J]. Holzforschung, 47 (2): 103-108.

FUJIMOTO T, KOBORI H, TSUCHIKAWA S. 2012. Prediction of wood density independently of moisture conditions using near infrared spectroscopy [J]. Journal of Near Infrared Spectroscopy, 20 (3): 353-359.

GILANI M S, TINGAUT P, HEEB M, et al. 2014. Influence of moisture on the vibro-mechanical properties of bio-engineered wood [J]. Journal of Materials Science, 49 (22): 7679-7687.

GLASS S V, BOARDMAN C R, ZELINKA S L. 2017. Short hold times in dynamic vapor sorption measurements mischaracterize the equilibrium moisture content of wood [J]. Wood Science and Technology, 51 (2): 243-260.

GUEVARA R, MOSLEMI A A. 1983. Hygroexpansive and sorptive behavior of wood modified with propylene oxide and oligomeric diisocyanate [J]. Journal of Wood Chemistry and Technology, 3 (1): 95-114.

GUO X, WU Y, YAN N. 2017. Characterizing spatial distribution of the adsorbed water in wood cell wall of Ginkgo biloba L. by μ-FTIR and confocal Raman spectroscopy [J]. Holzforschung, 71 (5): 415-423.

HALL L D, RAJANAYAGAM V. 1986. Evaluation of the distribution of water in wood by use of three dimensional proton NMR volume imaging [J]. Wood Science and Technology, 20 (4): 329-333.

HILL C A S, NORTON A J, NEWMAN G. 2010. The water vapor sorption properties of Sitka spruce determined using a dynamic vapor sorption apparatus [J]. Wood Science and Technology, 44 (3): 497-514.

HILL C A S, RAMSAY J, KEATING B, et al. 2012. The water vapor sorption properties of thermally modified and densified wood [J]. Journal of Materials Science, 47 (7): 3191-3197.

HOFSTETTER K, HINTERSTOISSER B, SALMEN L. 2006. Moisture uptake in native cellulose- the roles of different hydrogen bonds: a dynamic FTIR study using deuterium exchange [J]. Cellulose, 13 (2): 131-145.

HOGAN JR C J, NIKLAS K J. 2004. Temperature and water content effects on the viscoelastic behavior of Tilia americana(Tiliaceae) sapwood [J]. Trees, 18 (3): 339-345.

HOSSEINAEI O, WANG S, TAYLOR A M, et al. 2012. Effect of hemicellulose extraction on water absorption and mold susceptibility of wood-plastic composites [J]. International Biodeterioration and Biodegradation, 71: 29-35.

JALALUDIN Z, HILL C A S, SAMSI H W, et al. 2010. Analysis of water vapor sorption of oleo- thermal modified wood of acacia mangium and endospermum malaccense by a parallel exponential kinetics model and according to the Hailwood-Horrobin model [J]. Holzforschung, 64 (6): 763-770.

JOERDENS C, WIETZKE S, SCHELLER M, et al. 2010. Investigation of the water absorption in polyamide and wood plastic composite by terahertz time-domain spectroscopy [J]. Polymer Testing, 29 (2): 209-215.

KHAN M A, ALI K M I, WANG W. 1991. Electrical properties and X-ray diffraction of wood and wood plastic composite (WPC) [J]. International Journal of Radiation Applications and Instrumentation. Part C. Radiation Physics and Chemistry, 38 (3): 303-306.

KONDO T. 1997. The assignment of IR absorption bands due to free hydroxyl groups in cellulose [J]. Cellulose, 4 (4): 281-292.

KOUMBI-MOUNANGA T, GROVES K, LEBLON B, et al. 2015. Estimation of moisture content of trembling aspen (Populus tremuloides Michx.) strands by near infrared spectroscopy (NIRS) [J]. European Journal of Wood and Wood Products, 73 (1): 43-50.

LEBLON B, ADEDIPE O, HANS G, et al. 2013. A review of near-infrared spectroscopy for monitoring moisture content and density of solid wood [J]. Forestry Chronicle, 89 (5): 595-606.

LESTANDER T A, GELADI P. 2003. NIR spectroscopic measurement of moisture content in Scots pine seeds [J]. Analyst, 128 (4): 389-396.

MA E N, ZHAO G J, CAO J Z. 2005. Hygroexpansion of wood during moisture adsorption and desorption processes [J]. Forestry Studies in China, 7 (2): 43-46.

MAEDA H, FUKADA E. 1987. Effect of bound water on piezoelectric, dielectric, and elastic properties of wood [J]. Journal of Applied Polymer Science, 33 (4): 1187-1198.

MENON R S, MACKAY A L, FLIBOTTE S, et al. 1989. Quantitative separation of NMR images of water in wood on the basis of T_2 [J]. Journal of Magnetic Resonance, 82 (1): 205-210.

MERAKEB S, DUBOIS F, PETIT C. 2009. Modeling of the sorption hysteresis for wood [J]. Wood Science and Technology, 43 (7-8): 575-589.

MOLIŃSKI W, RACZKOWSKI J. 1988. Mechanical stresses generated by water adsorption in wood and their determination by tension creep measurements [J]. Wood Science and Technology, 22 (3): 193-198.

MORA C R, SCHIMLECK L R, YOON S, et al. 2011. Determination of basic density and moisture content of loblolly pine wood disks using a near infrared hyperspectral imaging system [J]. Journal of Near Infrared Spectroscopy, 19 (5): 401-409.

MOREIRA J L, SANTOS L. 2004. Spectroscopic interferences in fourier transform infrared wine analysis [J]. Analytica Chimica Acta, 513 (1): 263-268.

MOUDGIL K D, RAO D N, NARANG B S. 1985. Nuclear magnetic resonance and its applications in medicine. [J]. Indian Journal of Pediatrics, 52 (416): 231-241.

MURPHY D, PINHO M N D. 1995. An ATR-FTIR study of water in cellulose acetate membranes prepared by phase inversion [J]. Journal of Membrane Science, 106 (3): 245-257.

NAKANO T. 2006. Analysis of the temperature dependence of water sorption for wood on the basis of dual mode theory [J]. Journal of Wood Science, 52 (6): 490-495.

NASCIMBEM L B L R, RUBINI B R, POPPI R J. 2013. Determination of quality parameters in moist wood chips by near infrared spectroscopy combining PLS-DA and support vector machines [J]. Journal of Wood Chemistry and Technology, 33 (4): 247-257.

OBATAYA E, NORIMOTO M, GRIL J. 1998. The effects of adsorbed water on dynamic mechanical properties of wood. [J]. Polymer, 39 (14): 3059-3064.

OLSSON A M, SALMEN L. 2004. The association of water to cellulose and hemicellulose in paper examined by FTIR spectroscopy [J]. Carbohydrate Research, 339 (4): 813-818.

OZYHAR T, HERING S, SANABRIA S J, et al. 2013. Determining moisture-dependent elastic characteristics of beech wood by means of ultrasonic waves [J]. Wood Science and Technology, 47 (2): 329-341.

PERALTA P N. 1995. Modelling wood moisture sorption hysteresis using the independent-domain theory [J]. Wood and Fiber Science, 27 (3): 250-257.

PING Z H, NGUYEN Q T, CHEN S M, et al. 2001. States of water in different hydrophilic polymers-DSC and FTIR studies [J]. Polymer, 42 (20): 8461-8467.

POLETTO M, ZATTERA A J, SANTANA R M C. 2012. Structural differences between wood species: Evidence from chemical composition, FTIR spectroscopy, and thermogravimetric analysis [J]. Journal of Applied Polymer Science, 126 (S1): E337-E344.

POPESCU C, HILL C A S, CURLING S, et al. 2014. The water vapor sorption behavior of acetylated birch wood: How acetylation affects the sorption isotherm and accessible hydroxyl content [J]. Journal of Materials Science, 49 (5): 2362-2371.

RAUTKARI L, HILL C A S, CURLING S, et al. 2013. What is the role of the accessibility of wood hydroxyl groups in controlling moisture content? [J]. Journal of Materials Science, 48 (18): 6352-6356.

RIGGIN M T, SHARP A R, KAISER R, et al. 1979. Transverse NMR relaxation of water in wood [J]. Journal of Applied Polymer Science, 23 (11): 3147-3154.

SOMARATHNA H M C C, RAMAN S N, BADRI K H, et al. 2016. Quasi-static behavior of palm-based elastomeric polyurethane: for strengthening application of structures under impulsive loadings [J]. Polymers, 8 (5): 202.

SONG Y, WANG Z, YAN N, et al. 2016. Demethylation of wheat straw alkali lignin for application in phenol formaldehyde adhesives [J]. Polymers, 8 (6): 209.

SRINIVAS K, PANDEY K K. 2012. Effect of heat treatment on color changes, dimensional stability, and mechanical properties of wood [J]. Journal of Wood Chemistry and Technology, 32 (4): 304-316.

TAMRAKAR S, LOPEZ-ANIDO R A. 2011. Water absorption of wood polypropylene composite sheet piles and its influence on mechanical properties [J]. Construction and Building Materials, 25 (10): 3977-3988.

TANIGUCHI T, HARADA H, NAKATO K. 1978. Determination of water adsorption sites in wood by a hydrogen-deuterium exchange [J]. Nature, 272 (5650): 230-231.

THYGESEN L G, LUNDQVIST S. 2000. NIR measurement of moisture content in wood under unstable temperature conditions. Part 1. thermal effects in near infrared spectra of wood [J]. Journal of Near Infrared Spectroscopy, 8 (3): 183-189.

TIAN Q, KRAKOVSKY I, YAN G, et al. 2016. Microstructure changes in polyester polyurethane upon thermal and humid aging [J]. Polymers, 8 (5): 197.

WANG S Y, WANG H L. 1999. Effects of moisture content and specific gravity on static bending properties and hardness of six wood species [J]. Journal of Wood Science, 45 (2): 127-133.

XIE Y, HILL C A S, XIAO Z, et al. 2011. Dynamic water vapor sorption properties of wood treated with glutaraldehyde [J]. Wood Science and Technology, 45 (1): 49-61.

XIE Y, HILL C A S, XIAO Z, et al. 2010. Water vapor sorption kinetics of wood modified with glutaraldehyde [J]. Journal of Applied Polymer Science, 117 (3): 1674-1682.

XU F, YU J, TESSO T, et al. 2013. Qualitative and quantitative analysis of lignocellulosic biomass using infrared techniques: a mini-review [J]. Applied Energy, 104: 801-809.

ZAIHAN J, HILL C A S, CURLING S, et al. 2009. Moisture adsorption isotherms of acacia mangium and endospermum malaccense using dynamic vapor sorption [J]. Journal of Tropical Forest Science, 21 (3): 277-285.

ZHANG K, YU Q, ZHU L, et al. 2017. The preparations and water vapor barrier properties of polyimide films containing amide moieties [J]. Polymers, 9 (12): E677.

ZHANG M H. 2011. Mechanism of water sorption during adsorption process of wood studied by NMR [J]. Chinese Journal of Magnetic Resonance, 28 (1): 135-141.

ZHANG X, KUNZEL H M, ZILLIG W, et al. 2016. A Fickian model for temperature-dependent sorption hysteresis in hygrothermal modeling of wood materials [J]. International Journal of Heat and Mass Transfer, 100: 58-64.

Chapter 7   Molecular association of water with heat-treated wood cell walls during moisture adsorption process examined by micro-FTIR spectroscopy

7.1   Introduction

　　As a kind of engineering material, wood has great potential and has been used by human beings for thousands of years. From ancient houses, bridges, ships, paper and so on, to modern packaging materials, biological materials, fire-retardant materials, art materials, etc., have been used in wood. However, with the rapid increase of global population and sharp decrease of forest resources, the imbalance between supply and demand of resources is becoming more prominent. Our country is a large population country in the world, the demand for wood resources is very large, but at the same time our forest resources are also the world's poorest countries. In recent years, the imbalance of wood supply and demand is particularly prominent.

　　Wood is a kind of renewable and environmentally friendly natural polymer material with excellent mechanical properties, high strength-to-weight ratio and reasonable price, which is widely used in structural and non-structural materials. At the same time, wood also has hygroscopicity (Shaw and Elsken 1950; Zhang and Datta 2004; Zhang et al. 2016), dry shrinkage and swelling, which is mainly determined by the porous structure of wood and cellulose, hemicellulose and aromatic polymer lignin properties of the components of the cell (Guindos 2014; Hanninen et al. 2011; Hozjan and Svensson 2011; Terashima et al. 2009). It is very difficult for water molecule to enter the crystallization region of wood cellulose, while hemicellulose and cellulose in uncrystallized region have strong water affinity. The hygroscopicity of wood with the change of environmental temperature and humidity causes the displacement among the amorphous region of cellulose, hemicellulose and lignin molecules, and the displacement and interlacing between the cell walls caused by the movement of these chemical components, which results in the formation of wood defects, such as radial splitting and cracks, and has a negative impact on the efficient utilization of wood.

The lack of forest resources has become an indisputable fact, so it is very important to strictly control the quality of wood raw materials, so as to ensure the production of high-quality wood products, but also to ensure the efficient and rational use of wood resources. In order to improve various properties of wood, many methods to modify the chemical structure of wood have been introduced by researchers (Bami and Mohebby 2011; Donath et al. 2004; Esteves and Pereira 2009; Kartal et al. 2007; Torgovnikov and Vinden 2010), for example, chemical modification, heat treatment modification, enzymatic catalysis modification, pressure impregnation modification. These methods improve the shrinkage and swelling, poor dimensional stability, discoloration, flammability and non- corrosion resistance of wood to varying degrees. At the same time, they give the wood specific color, specific physical and chemical properties and other characteristics, and make the wood higher grade, thus expanding its application scope and increasing its service life. There are a lot of reports on the improvement of wood physicochemical properties by heat treatment in foreign countries, and there are several mature treatment processes and related equipment. However, there are few reports in our country. The heat treatment process is still in its infancy in our country. Heat treatment has many advantages; for example, through heat treatment, it can effectively reduce the moisture absorption of wood (Borrega and Karenlampi 2010; Joma et al. 2016; Modes et al. 2013; Scheiding et al. 2016; Toubal et al. 2016), wood size stability, durability (Altgen et al. 2016; Bekhta and Niemz 2003; Rowell et al. 2009; Srinivas and Pandey 2012), thus extending the carbon fixation function of wood to improve the ecological environment, wood color changes throughout, and can imitate valuable species wood color. Meanwhile, the heat treatment method avoids the pollution caused by the use of chemical agents. The wood obtained by heat treatment is non-toxic and pollution-free and belongs to environment-friendly material.

Therefore, it can be seen from the advantages of wood heat treatment, water adsorption characteristics are prime properties of heat treated wood, and fundamental understanding of the water adsorption of heat treated wood would be highly beneficial for the utilization of this material.

There exist lots of publications available which have provided useful information about water adsorption of heat-treated wood (Bastani et al. 2015; Kartal et al. 2007; Metsä-Kortelainen and Viitanen 2012; Scheiding et al. 2016; Tomak et al. 2011; Wang et al. 2011; Willems et al. 2015). For example, Kartal *et al.* showed the heat-treated Sugi (*Cryptomeria japonica* D. Don) sapwood adsorbed less water than unheated specimens (Kartal et al. 2007). Meanwhile, Metsä-Kortelainen *et al.* confirmed that heat modification evidently reduced the hygroscopicity of Scots pine (*Pinus sylvestris*) and Norway spruce (*Picea abies*) (Metsä-Kortelainen et al. 2006). Furthermore, the change of hygroscopicity is attributed to chemical modification of wood (Windeisen et al. 2007). The chemical modification of heat-treated wood during heat treatment is analyzed by different chemical analyzing methods, and similar results are reported and discussed (Gonzalez-Pena et al. 2009; Huang et al. 2013; Ozgenc et al. 2017; Tjeerdsma and Militz 2005). Hakkou *et al.* proved that heat treatment could lead to conformational modification of polysaccharide components and reorganization of lignocellulosic polymeric components (Hakkou et al. 2005). Boonstra *et al.* stated that the hygroscopicity was reduced because of the cross-linking of lignin and the decreasing of OH groups (Boonstra and Tjeerdsma 2006). Although the water adsorption of heat-treated wood is characterized and the decrease of hygroscopicity is due to chemical modification of wood components, until recently there has been a lack of understanding about existing states of water in heat-treated wood, which is evidently an important part of the water adsorption mechanism of heat-treated wood.

The water in wood exists in the capillary system composed of the cell cavity, the cell gap and the cell wall gap. There are three kinds of classifications of water in wood (Araujo et al. 1992; Araujo et al. 1993; Gezici-Koc et al. 2017; Menon et al. 1987), one is free water in the lumen and intercellular space, the other is adsorbed water in the cellulose layer of the cell wall, and the third is combined water which forms the polymer compound of wood fiber. Among them, free water and wood are physically combined, not tightly combined, and are easy to escape from wood. During wood drying process, the first evaporation is free water, the increase and decrease of wood mechanics and other properties have no effect, only the weight of wood, electrical and thermal properties. Adsorbed water and wood chemical composition is physico-chemical combination adsorbed water from wood which is not easy to discharge; only in the free water evaporation, wood vapor pressure is greater than the air vapor pressure, which can evaporate. The change of adsorbed water is the turning point of wood properties, and all properties of wood change with the increase or decrease of adsorbed water. Although the water content in wood has been known to some extent, how to characterize the states of water in wood is still an active research field, and there is still no answer to the question (Berthold et al. 1996). Many techniques such as near-infrared spectroscopy (Agarwal and Kawai 2005;

Berthold et al. 1998; Inagaki et al. 2008; Lestander 2008; Mora et al. 2011), dynamic vapor sorption (DVS) (Glass et al. 2018; Hill et al. 2010; Sharratt et al. 2010; Xie et al. 2011; Zaihan et al. 2009), nuclear magnetic resonance (Araujo et al. 1992; Casieri et al. 2004; Kekkonen et al. 2014; Menon et al. 1989; Menon et al. 1987; Vogt et al. 2004), FTIR spectroscopy (Almgren et al. 2008; Celino et al. 2014; Hakkou et al. 2005) and Raman spectroscopy (Agarwal and Kawai 2005; Ding et al. 2016) have been employed to get a better understanding about the existing states of water in wood. Among these techniques, FTIR spectroscopy has these outstanding advantages of being highly sensitive to water and being able to characterize minor structural changes of chemical components (Akerholm and Salmen 2004; Poletto et al. 2012; Xie et al. 2015). With the progress of science and technology, more advanced FTIR technique such as micro-FTIR spectrometer has been introduced which equipped with an additional visible-light microscope can be used to visualize morphology and select the observation area of a micro-sized sample and own a better sensitivity in the detection of molecular structures. Liu *et al.* showed that micro-FTIR spectroscopy was relatively simple, inexpensive, and readily available for quantitatively investigating the water adsorption of micron-sized samples (Liu et al. 2008). Therefore, micro-FTIR spectroscopy is considered to be a promising tool for studying water adsorption of heat-treated wood.

Our objective in this study was to characterize the states of water in wood and demonstrate the molecular structure of adsorbed water in heat-treated wood. In this study, we first collected the spectra of untreated wood and heat-treated wood by micro-FTIR spectrometer, observed the chemical changes of heat-treated wood, and we used micro-FTIR spectrometer to collect *in situ* spectra of a typical heat-treated wood over a wide range of RH levels, with the aid of a specially designed sample chamber. These micro-FTIR spectra were applied to characterize the states and molecular structure of adsorbed water after a series of finishing and fitting.

7.2 Experimental section

7.2.1 Sample preparation

One deciduous tree, *Ginkgo biloba* L. (Ginkgoaceae), was selected in this study. Raw materials in terms of *ginkgo biloba* trees were got from Hebei province of China which were 5-to 7-year-old. The dimension of wood specimens was set 100 mm in length, 30 mm in width and 20 mm in thickness. Prior to the heat treatment, twelve replicate specimens were oven-dried at (103±2) ℃ until a constant weight was obtained as the dry weight before the heat-treatment. After cooling in a desiccator, these specimens were placed in the vacuum drying oven under the pressure of 5 Pa. The following heat-treatment procedure was employed as follows: (Ⅰ) six replicate specimens were heating from initial temperature [(25±2) ℃] to the set temperature (180 ℃) which was raised 20 ℃ every 10 min, (Ⅱ) specimens were kept at 180℃ for 4 h, and (Ⅲ) specimens were cooled until the temperature dropped below 100 ℃, and the cooling time lasted 5h. Meanwhile, the whole heat-treatment process took place in a low-oxygen environment. Afterwards, six heat-treated specimens were oven-dried at (103±2) ℃ again to obtain the dry weight after the heat treatment, and the mass loss of heat-treated wood was (3.8±0.5) %. Subsequently, transverse sections, 0.5 cm × 0.5cm × 10 μm, were prepared by the sliding microtome (Leica 2010R, Leica Microsystems, Wetzlar, Germany) from heat-treated specimens. And then, the transverse sections were placed into two glass slides for micro-FTIR measurements. At the same time, the same cross-section of the unheat-treated specimen was prepared and put into two glass slides for micro- FTIR measurements.

7.2.2 Micro-FTIR spectroscopy equipment

Figure 7.1 gives the micro-FTIR equipment used to study the molecular association of adsorbed water with thermo-treated wood. The micro-FTIR equipment is Nicolet IN 10^{TM} micro-FTIR spectrometer (Thermo Electron Scientific Instruments, Madison, WI, USA), which was equipped with microscope, dynamically aligned high-speed interferometer and mercury-cadmium-telluride (MCT) photo conductive detector. In this microscope, there were two pathways. One was for visible beam which could be used for visual examination and alternated orientation, the other was for IR beam on which the aperture was mounted for defining the observation area of sample. And the location and size of the observation area could be adjusted according to the requirements of specific experiments with a minimum size of 3 μm×3 μm through which IR beam can pass and be collected by MCT photoconductive detector. For using the dynamically aligned high-speed interferometer and MCT photoconductive detector, each spectrum could be recorded with 4 cm^{-1} spectral resolution between 720 cm^{-1} and 4000 cm^{-1}.

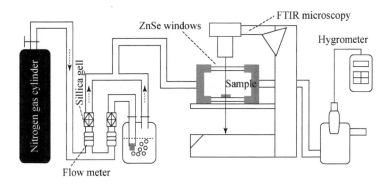

Figure 7.1　Schematic diagram of the experimental apparatus for micro-FTIR spectroscopy measurements

In order to accurately control the RH of sample, one specially designed sample cell was introduced. As shown in Figure 7.1, the sample cell was composed of two ZnSe plates with the thickness of 2 mm and sample was deposited on the lower plate. Meanwhile, the sample cell contained two pipes, one is for the stream of water-saturated N_2 and the other is for dry N_2. The RH in sample cell was adjusted by these mixing gases which were controlled by two flow rate meters (Alicat scientific, Tucson, AZ, USA). Near the exit point of the sample cell, there was one hygrometer (Center 310, Center Technology Corp., New Taipei, Taiwan, China) which was used to measure the resulting RH of sample cell. Before experiment, the sample cell was dried with continuous dry N_2 for 12 h. And then the RH was adjusted according to the requirements of specific experiments. When the RH reached at a desired value, a period of 60 min was maintained to allow the RH to be stabilized. Sample spectra and corresponding background spectra taken on a blank ZnSe substrate were measured successively at this desired selected RH.

7.2.3　Data processing

To gain information about the adsorbed water by thermo-treated wood from the detailed spectral change of micro-FTIR, FTIR subtractive spectroscopy technique was used, and then difference spectra between the measured spectra at various RH and the spectrum at 0 RH% were obtained in the OMINIC software. In analysis of detailed spectral change of difference spectra, the band decompositions were achieved by a Gaussian/lorentzian mixing method using the curve fitting routine provided in the OriginPro software package. The Gaussian/lorentzian mixing function was as follows.

$$y = y_0 + A\left[m_u \cdot \frac{2}{\pi} \cdot \frac{w}{\left(4(x-x_0)^2 + w^2\right)} + (1-m_u) \cdot \frac{\sqrt{4\ln 2}}{\sqrt{\pi}w} \cdot e^{-4\ln 2/w^2 \cdot (x-x_0)^2} \right] \quad (7.1)$$

In this function, x_0 is the peak position, y_0 is the baseline offset, A is the total area, m_u is the profile shape factor and w is the full width of the peak at half maximum (FWHM). In the fitting process, x_0 was kept unchanged, and all other parameters were allowed to vary upon iteration.

7.3 Results and discussion

7.3.1 FTIR spectra of the heat-treated wood associated with water molecules

Figure 7.2 shows the micro-FTIR spectra of untreated wood and heat-treated wood at 180°C. Compared with the untreated wood, chemical changes were observed after heat treatment. For hemicellulose and lignin, the characteristic peaks at 1739 cm^{-1}, 1594 cm^{-1}, and 811 cm^{-1}, belonging to C=O stretching vibrations in the O=C—OH group of glucuronic acid unit in xylan, to the aromatic skeletal vibrations plus C=O stretch of lignin, and to vibrations caused by the equatorially aligned hydrogen at the C_2 atom in the mannose residue of glucomannan, respectively (Hillis and Rozsa 1978; Zylka et al. 2009). In general, the relative intensities of the absorption peaks at 1739 cm^{-1}, 1594 cm^{-1} and 811 cm^{-1} decreased under heat treatment, indicating that the treatment led to degradation of C=O in the O=C—OH group of the glucuronic acid unit of xylan, a loss of the C=O group linked to the aromatic skeleton in lignin, as well as a decomposition of the glucomannan backbone. Meanwhile, the xylan band at 1452 cm^{-1} ascribed to CH_2 symmetric bending on the xylose ring and the peak at 1230 cm^{-1} belonging to the C=O stretching in the O=C—OH group showed obvious changes. The band 1594 cm^{-1} showed a larger decrease after heat treatment. The intensity changes were though as large for other absorption bands, 1507 cm^{-1} and 1264 cm^{-1} ascribed to the aromatic skeletal vibration and the vibration of the guaiacyl ring, together with the C=O stretch, respectively. The larger changes in the relative intensity at 1594 cm^{-1}, in combination with the lower change in intensity at 1507 cm^{-1}, indicate that a loss of the C=O group linked to the aromatic skeleton of lignin has probably occurred. This could indicate that cross-links have been formed between aromatic units in the lignin.

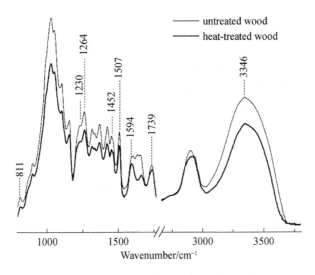

Figure 7.2 Micro-FTIR spectra of untreated wood and heat-treated wood

Figure 7.3 shows micro-FTIR spectra of heat-treated wood at various RH levels (from 0% to 92%). As the RH increased from 0% to 92%, the broad and strong band centered at 3346 cm^{-1} increased, showing that the water molecules were absorbed successively by O—H groups. Meanwhile, the characteristic peaks at 1739 cm^{-1} and 1594 cm^{-1} shifted to 1736 cm^{-1} and 1596 cm^{-1}, respectively. These indicated that carbonyl C=O and C—O groups preferentially combined with water molecules to form hydrogen bonds as the RH increased.

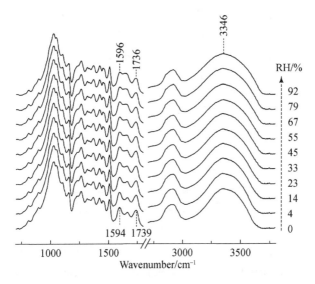

Figure 7.3 Micro-FTIR spectra of heat-treated wood at various RH levels

7.3.2 The analysis of difference spectra

Water adsorption in micro-FTIR spectroscopy provided only an unresolved broad band, which offered limited precise information. This study further aimed to extract information about minor structural changes occurring with respect to water adsorption, and thus difference spectra were employed, as described in the data processing section.

Figure 7.4 gives difference spectra of the heat-treated wood adsorbed by water at various RH (from 4% to 92%). As the RH increased, the intensity in the wavenumber region 2900-3700 cm^{-1} was found to increase obviously showing that the heat treated wood still had water adsorption capacity. It is consistent with the fact that the heat treated wood was also hygroscopic material although heat treatment can effectively reduce the moisture adsorption of wood. During the moisture adsorption of heat treated wood, the bands at 1755 cm^{-1} were observed to decrease in intensity with increase in RH over the full range from 4% to 92%, while the peaks at 1257 cm^{-1} increase. The band at 1755 cm^{-1} was identified to be the free carbonyl C=O groups. Likewise, the bands at 1257 were assigned to the hydrogen bonded carbonyl C—O group. These negative-going band of 1755 cm^{-1} and positive-going band at 1257 cm^{-1} changed continuously upon moisture adsorption, indicating that the carbonyl C=O and C—O group were active sites for water adsorption. Meanwhile, the band at 1171 cm^{-1}

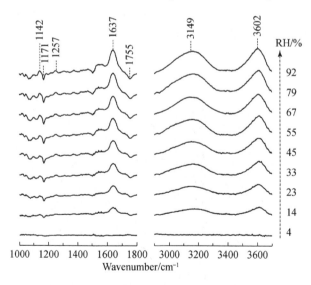

Figure 7.4 Micro-FTIR difference spectra at various RH levels after subtraction of the spectrum taken at the lowest RH (0%). The arrow means the RH is increasing from 4% to 92%

was observed to decrease in intensity with increase in RH. However, the intensity of the band at 1142 cm^{-1} was observed to rise. These negative-going and positive-going was due to the effect of water adsorption on O(3)H···O(5) hydrogen bond.

7.3.3 Intermolecular interactions between adsorbed water and the heat-treated wood

It is well known that the band shape and position can give indication of the presence of more than one component, so the broad band at 2900-3700 cm^{-1} in Figure 7.5 was obviously composed of several component peaks. In order to differentiate varieties of component peaks for demonstrating intermolecular interactions between adsorbed water and the heat treated wood, the fitting process which was described in the experimental section was applied to the spectral region of 2900-3700 cm^{-1}.

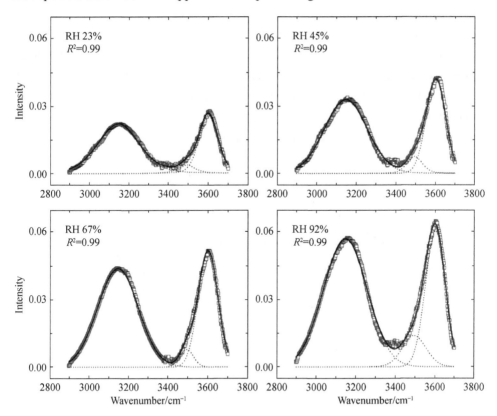

Figure 7.5   Fitted band contours of micro-FTIR difference spectra in the region of 2900-3700 cm^{-1}

Square dot: the experimental micro-FTIR difference spectra; Dashed lines: three fitted spectral bands; Solid lines: the sum spectra of the three fitted spectral components; R^2: the coefficient of determination

The difference spectra, band-fitting results and corresponding component bands are displayed in Figure 7.5. These corresponding component bands were positioned at 3149 cm^{-1}, 3496 cm^{-1} and 3602 cm^{-1}, respectively, whose frequencies were similar to those obtained by Olsson et al. These frequencies were highly sensitive to hydrogen bonding. As the hydrogen bonding interactions decreased, a strengthening of the O — H oscillator resulted in a blue shift in energy. The strong spectral intensity at lower energies (the fitting peak at 3149 cm^{-1}) was the characteristic of strongly hydrogen-bonded water molecules. The spectral intensity at moderate energies (the fitting peak at 3496cm^{-1}) was assigned to moderately hydrogen-bonded water molecules. In contrast, the intensity at higher energies (3602cm^{-1}) was the charact eristic of weakly hydrogen-bonded water molecules.

In order to illustrate how the amount of the different hydrogen-bonded water molecules changed with RH, the integrated intensities of each component in RH are shown in Figure 7.6. As the RH increased, the content of each component had different change. According to the curve trend with the RH in Figure 7.6, three sections (I, II and III) were divided. For heat treated wood, the three divided sections existed obvious differences.

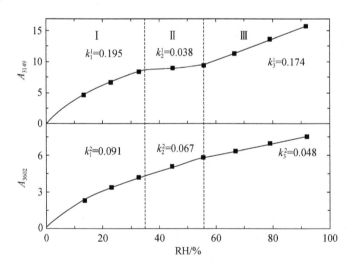

Figure 7.6　The integrated intensities of two fitted band components for the micro-FTIR difference spectra at various RH levels. k: the slope of the fitting curve

For the wood heat treated at 180℃, Section I was RH below 35%. In this section, both A_{3149} and A_{3602} increased obviously, while the growth trend of A_{3149} was higher than that of A_{3602} with the RH, meaning that the water uptake preferred to form strong hydrogen bonds in this region. It is known that strong hydrogen-bonded components should mainly come from the water bound directly by hydrogen bonds to hydrophilic groups of the cellulose and the hemicelluloses and the tetrahedral coordinated water molecules participating in strong hydrogen bonding interactions with adjacent molecules. Considering that the strongest hydrogen bonded H_2O was trapped at such low RHs, it can be suggested that the water molecules should have bound directly to the net of hydrophilic groups. As previously mentioned, for the heat-treated wood, the hydroxyl groups and carboxyl groups were the effective adsorption sites, which could be demonstrated in the molecular structures of cellulose, hemicellulose and lignin [Figure 7.7 (a)]. Thus, in this region the absorbed water molecules should have these structures as $C=O\cdots(HOH)\cdots OH$ and $OH\cdots(OH_2)\cdots OH$ as shown in Figure 7.7 (b).

As the RH was in the range of 35%–55% (section II), the A_{3149} almost remained constant. The growth trend of A_{3602} was obvious, showing that water uptake preferred to form the weak hydrogen-bonding interactions in this region. This behavior indicated that the first hydration layer of hydrophilic groups was almost fully occupied, and thus water molecules had to indirectly bond to these hydrophilic groups via another water molecule as shown in Figure 7.7 (c). When the RH was about 55% (section III), the continuous increasing RH led to A_{3149} and A_{3602} increasing, while the growth trend of A_{3149} was higher than that of A_{3602} with the RH, which shows that the increasing of strong hydrogen bonding component was faster than that of weak hydrogen bonding component. It is well known that the hydrogen-bonds in the five-molecule tetrahedral structure are strong, while those in other structures are weak. Therefore, at RH=55%, strong hydrogen-bonded components should mainly come from the five-molecule tetrahedral structure of water and the water interacted with themselves to form water clusters as shown in Figure 7.7 (d).

(a)

(b)

(c)

(c)

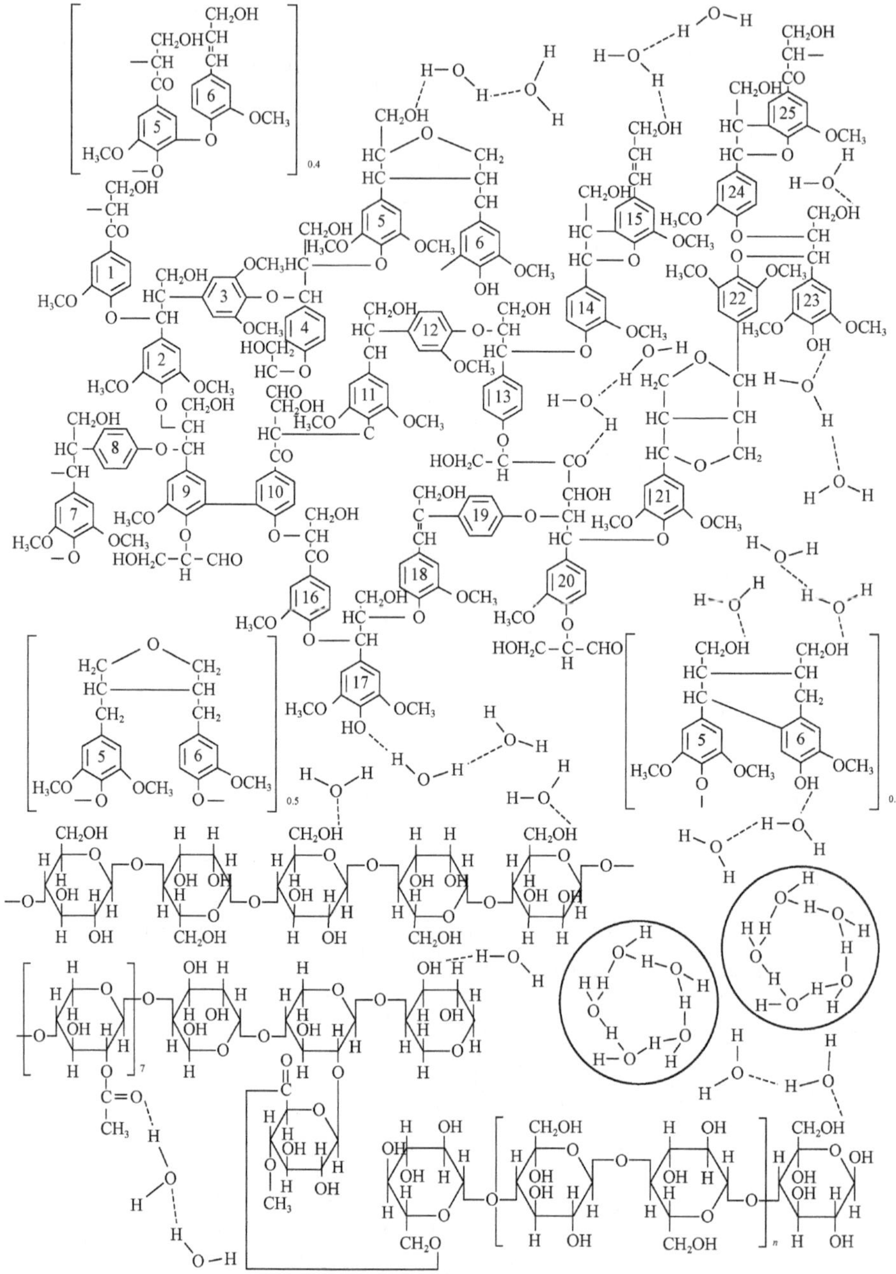

Figure 7.7 Proposed water adsorption mechanism and corresponding molecular structures at three sections (a) Molecular structures of cellulose, hemicellulose and lignin; (b) When the RH is lower than 35%, the water molecules prefer to form strong hydrogen bonds; (c) As the RH is in the range of 35%–55%, the water molecules prefer to form weak hydrogen bonds; (d) When the RH enters 55%, the water molecules interact with themselves to form water clusters

7.4 Conclusions

By using *in situ* micro-FTIR spectroscopy and a specially designed sample cell, a detailed investigation on molecular association of adsorbed water with the heat treated wood was achieved during adsorption process. For the micro-FTIR spectra of untreated and heat-treated wood, the relative intensities of the absorption peaks at 1739 cm^{-1}, 1594 cm^{-1} and 811 cm^{-1} indicated the degradation of $C=O$ in the $O=C-OH$ group of the glucuronic acid unit of xylan, a loss of the $C=O$ group linked to the aromatic skeleton in lignin, as well as a decomposition of the glucomannan backbone. For the micro-FTIR spectra of heat-treated wood at various RH levels, as the RH increased from 0% to 92%, the characteristic peaks at 1739 cm^{-1} and 1594 cm^{-1} exhibited a shift, showing that carboxyl $C=O$ and $C-O$ groups were active sites for water adsorption. Meanwhile, the different spectra between the humid and dry spectra of heat treated wood confirmed carbonyl $C=O$ and $C-O$ groups combined with water molecules to form hydrogen bonds. From component band analysis of the spectral range of 2900-3700 cm^{-1}, three peaks at 3149 cm^{-1}, 3496 cm^{-1} and 3602 cm^{-1} were identified and assigned to strongly, moderately and weakly hydrogen-bonded water molecules, respectively. As expected, the content of three components performed different change, and then three sections (Ⅰ, Ⅱ and Ⅲ) were divided according to the variation trend. For the wood heat treatment at 180℃, Section Ⅰ was below 35% RH, Section Ⅱ was in the range of 35%-55%, and Section Ⅲ was higher than 55%.

References

AGARWAL U P, KAWAI N. 2005. "Self-absorption" phenomenon in near-infrared Fourier transform Raman spectroscopy of cellulosic and lignocellulosic materials [J]. Applied Spectroscopy, 59 (3): 385-388.

AKERHOLM M, SALMEN L. 2004. Softening of wood polymers induced by moisture studied by dynamic FTIR spectroscopy [J]. Journal of Applied Polymer Science, 94 (5): 2032-2040.

ALMGREN K M, AKERHOLM M, GAMSTEDT E K, et al. 2008. Effects of moisture on dynamic mechanical properties of wood fiber composites studied by dynamic FTIR spectroscopy [J]. Journal of Reinforced Plastics and Composites, 27 (16-17): 1709-1721.

ALTGEN M, HOFMANN T, MILITZ H. 2016. Wood moisture content during the thermal modification process affects the improvement in hygroscopicity of Scots pine sapwood [J]. Wood Science and Technology, 50 (6): 1181-1195.

ARAUJO C D, MACKAY A L, HAILEY J, et al. 1992. Proton magnetic-resonance techniques for characterization of water in wood - application to white spruce [J]. Wood Science and Technology, 26 (2): 101-113.

ARAUJO C D, MACKAY A L, WHITTALL K P, et al. 1993. A diffusion-model for spin-spin relaxation of compartmentalized water in wood [J]. Journal of Magnetic Resonance Series B, 101 (3): 248-261.

BAMI L K, MOHEBBY B. 2011. Bioresistance of poplar wood compressed by combined hydro-thermo-mechanical wood modification (CHTM): soft rot and brown-rot [J]. International Biodeterioration & Biodegradation, 65 (6): 866-870.

BASTANI A, ADAMOPOULOS S, MILITZ H. 2015. Water uptake and wetting behavior of furfurylated, N-methylol melamine modified and heat-treated wood [J]. European Journal of Wood and Wood Products, 73 (5): 627-634.

BEKHTA P, NIEMZ P. 2003. Effect of high temperature on the change in color, dimensional stability and mechanical properties of spruce wood [J]. Holzforschung, 57 (5): 539-546.

BERTHOLD J, OLSSON R, SALMEN L. 1998. Water sorption to hydroxyl and carboxylic acid groups in Carboxymethylcellulose (CMC) studied with NIR-spectroscopy [J]. Cellulose, 5 (4): 281-298.

BERTHOLD J, RINAUDO M, SALMEN L. 1996. Association of water to polar groups; Estimations by an adsorption model for ligno-cellulosic materials [J]. Colloids and Surfaces A-Physicochemical and Engineering Aspects, 112 (2-3): 117-129.

BOONSTRA M J, TJEERDSMA B. 2006. Chemical analysis of heat treated softwoods [J]. Holz Als Roh-Und Werkstoff, 64 (3): 204-211.

BORREGA M, KARENLAMPI P P. 2010. Hygroscopicity of heat treatment Norway spruce (Picea abies) wood [J]. European Journal of Wood and Wood Products, 68 (2): 233-235.

CASIERI C, SENNI L, ROMAGNOLI M, et al. 2004. Determination of moisture fraction in wood by mobile NMR device [J]. Journal of Magnetic Resonance, 171 (2): 364-372.

CELINO A, GONCALVES O, JACQUEMIN F et al. 2014. Qualitative and quantitative assessment of water sorption in natural fibers using ATR-FTIR spectroscopy [J]. Carbohydrate Polymers, 101: 163-170.

DING T, WANG C, PENG W. 2016. A theoretical study of moisture sorption behavior of heat-treated pine wood using Raman spectroscopic analysis [J]. Journal of Forestry Engineering, 1 (5): 15-19.

DONATH S, MILITZ H, MAI C. 2004. Wood modification with alkoxysilanes [J]. Wood Science and Technology, 38 (7): 555-566.

ESTEVES B M, PEREIRA H M. 2009. Wood modification by heat treatment: A review [J]. BioResources, 4 (1): 370-404.

GEZICI-KOC O, ERICH S J F, HUININK H P, et al. 2017. Bound and free water distribution in wood during water uptake and drying as measured by 1D magnetic resonance imaging [J]. Cellulose, 24 (2): 535-553.

GLASS S V, BOARDMAN C R, THYBRING E E, et al. 2018. Quantifying and reducing errors in equilibrium moisture content measurements with dynamic vapor sorption (DVS) experiments [J]. Wood Science and Technology, 52 (4): 909-927.

GONZALEZ-PENA M M, CURLING S F, HALE M D C. 2009. On the effect of heat on the chemical composition and dimensions of thermally-modified wood [J]. Polymer Degradation and Stability, 94 (12): 2184-2193.

GUINDOS P. 2014. Numerical modeling of timber with knots: the progressively damaged lattice approach vs. The equivalent damaged continuum [J]. Holzforschung, 68 (5): 599-613.

HAKKOU M, PETRISSANS M, ZOULALIAN A, et al. 2005. Investigation of wood wettability changes during heat treatment on the basis of chemical analysis [J]. Polymer Degradation and Stability, 89 (1): 1-5.

HANNINEN T, KONTTURI E, VUORINEN T. 2011. Distribution of lignin and its coniferyl alcohol and coniferyl aldehyde groups in Picea abies and Pinus sylvestris as observed by Raman imaging [J]. Phytochemistry, 72 (14-15): 1889-1895.

HILL C A S, NORTON A J, NEWMAN G. 2010. The water vapor sorption properties of Sitka spruce determined using a dynamic vapor sorption apparatus [J]. Wood Science and Technology, 44 (3): 497-514.

HILLIS W E, ROZSA A N. 1978. Softening temperatures of wood [J]. Holzforschung, 32 (2): 68-73.

HOZJAN T, SVENSSON S. 2011. Theoretical analysis of moisture transport in wood as an open porous hygroscopic material [J]. Holzforschung, 65 (1): 97-102.

HUANG X, KOCAEFE D, KOCAEFE Y, et al. 2013. Structural analysis of heat-treated birch (Betule papyrifera) surface during artificial weathering [J]. Applied Surface Science, 264: 117-127.

INAGAKI T, YONENOBU H, TSUCHIKAWA S. 2008. Near-infrared spectroscopic monitoring of the water adsorption/desorption process in modern and archaeological wood [J]. Applied Spectroscopy, 62 (8): 860-865.

JOMA E, SCHMIDT G, GONÇALVES CREMONEZ V, et al. 2016. The effect of heat treatment on Wood-Water relationship and mechanical properties of commercial uruguayan plantation timber eucalyptus grandis [J]. Australian Journal of Basic and Applied Sciences, 10 (1): 704-708.

KARTAL S N, HWANG W, IMAMURA Y. 2007. Water absorption of boron-treated and heat-modified wood [J]. Journal of Wood Science, 53 (5): 454-457.

KARTAL S N, HWANG W, YAMAMOTO A, et al. 2007. Wood modification with a commercial silicon emulsion: Effects on boron release and decay and termite resistance [J]. International Biodeterioration & Biodegradation, 60 (3): 189-196.

KEKKONEN P M, YLISASSI A, TELKKI V. 2014. Absorption of water in thermally modified pine wood as studied by nuclear magnetic resonance [J]. Journal of Physical Chemistry C, 118 (4): 2146-2153.

LESTANDER T A. 2008. Water absorption thermodynamics in single wood pellets modelled by multivariate near-infrared spectroscopy [J]. Holzforschung, 62 (4): 429-434.

LIU Y, YANG Z, DESYATERIK Y, et al. 2008. Hygroscopic behavior of substrate-deposited particles studied by micro-FTIR spectroscopy and complementary methods of particle analysis [J]. Analytical Chemistry, 80 (3): 633-642.

MENON R S, MACKAY A L, FLIBOTTE S, et al. 1989. Quantitative separation of NMR images of water in wood on the basis of T2 [J]. Journal of Magnetic Resonance, 82 (1): 205-210.

MENON R S, MACKAY A L, HAILEY J, et al. 1987. An NMR determination of the physiological water distribution in wood during drying [J]. Journal of Applied Polymer Science, 33 (4): 1141-1155.

METSÄ-KORTELAINEN S, ANTIKAINEN T, VIITANIEMI P. 2006. The water absorption of sapwood and heartwood of Scots pine and Norway spruce heat-treated at 170 ℃, 190 ℃, 210 ℃ and 230 ℃ [J]. Holz Als Roh-Und Werkstoff, 64 (3): 192-197.

METSÄ-KORTELAINEN S, VIITANEN H. 2012. Wettability of sapwood and heartwood of thermally modified Norway spruce and Scots pine [J]. European Journal of Wood and Wood Products, 70 (1-3): 135-139.

MODES K S, SANTINI E J, VIVIAN M A. 2013. Hygroscopicity of wood from eucalyptus grandis and Pinus taeda subjected to thermal treatment [J]. Cerne, 19 (1): 19-25.

MORA C R, SCHIMLECK L R, YOON S, et al. 2011. Determination of basic density and moisture content of loblolly pine wood disks using a near infrared hyperspectral imaging system [J]. Journal of Near Infrared Spectroscopy, 19 (5SI): 401-409.

OZGENC O, DURMAZ S, BOYACI I H, et al. 2017. Determination of chemical changes in heat-treated wood using ATR-FTIR and FT Raman spectrometry [J]. Spectrochimica Acta Part A-Molecular and Biomolecular Spectroscopy, 171: 395-400.

POLETTO M, ZATTERA A J, SANTANA R M C. 2012. Structural differences between wood species: Evidence from chemical composition, FTIR spectroscopy, and thermogravimetric analysis [J]. Journal of Applied Polymer Science, 1261 (SI): E336-E343.

ROWELL R M, IBACH R E, MCSWEENY J D, et al. 2009. Understanding decay resistance, dimensional stability and strength changes in heat treated and acetylated wood [J]. Wood Material Science and Engineering, 4 (1-2): 14-22.

SCHEIDING W, DIRESKE M, ZAUER M. 2016. Water absorption of untreated and thermally modified sapwood and heartwood of Pinus sylvestris L. [J]. European Journal of Wood and Wood Products, 74 (4): 585-589.

SHARRATT V, HILL C A S, ZAIHAN J, et al. 2010. Photodegradation and weathering effects on timber surface moisture profiles as studied using dynamic vapor sorption [J]. Polymer Degradation and Stability, 95 (12): 2659-2662.

SHAW T M, ELSKEN R H. 1950. Nuclear magnetic resonance absorption in hygroscopic materials [J]. Journal of Chemical Physics, 18 (8): 1113-1114.

SRINIVAS K, PANDEY K K. 2012. Effect of heat treatment on color changes, dimensional stability, and mechanical properties of wood [J]. Journal of Wood Chemistry and Technology, 32 (4): 304-316.

TERASHIMA N, KITANO K, KOJIMA M, et al. 2009. Nanostructural assembly of cellulose, hemicellulose, and lignin in the middle layer of secondary wall of ginkgo tracheid [J]. Journal of Wood Science, 55 (6SI): 409-416.

TJEERDSMA B F, MILITZ H. 2005. Chemical changes in hydrothermal treated wood: FTIR analysis of combined hydrothermal and dry heat-treated wood [J]. Holz als Roh-Und Werkstoff, 63 (2): 102-111.

TOMAK E D, VIITANEN H, YILDIZ U C, et al. 2011. The combined effects of boron and oil heat treatment on the properties of beech and Scots pine wood. Part 2: water absorption, compression strength, color changes, and decay resistance [J]. Journal of Materials Science, 46 (3): 608-615.

TORGOVNIKOV G, VINDEN P. 2010. Microwave wood modification technology and its applications [J]. Forest Products Journal, 60 (2): 173-182.

TOUBAL L, CUILLIERE J, BENSALEM K, et al. 2016. Hygrothermal effect on moisture kinetics and mechanical properties of Hemp/Polypropylene composite: Experimental and numerical studies [J]. Polymer Composites, 37 (8): 2342-2352.

VOGT B D, SOLES C L, LEE H J, et al. 2004. Moisture absorption and absorption kinetics in polyelectrolyte films: influence of film thickness [J]. Langmuir, 20 (4): 1453-1458.

WANG S, MAHLBERG R, JAMSA S, et al. 2011. Surface properties and moisture behavior of pine and heat-treated spruce modified with alkoxysilanes by sol-gel process [J]. Progress in Organic Coatings, 71 (3): 274-282.

WILLEMS W, ALTGEN M, MILITZ H. 2015. Comparison of EMC and durability of heat treated wood from high versus low water vapor pressure reactor systems [J]. International Wood Products Journal, 6 (1): 21-26.

WINDEISEN E, STROBEL C, WEGENER G. 2007. Chemical changes during the production of thermo-treated beech wood [J]. Wood Science and Technology, 41 (6): 523-536.

XIE J, QI J, HUANG X, et al. 2015. Comparative analysis of modern and ancient buried Phoebe zhennan wood: Surface color, chemical components, infrared spectroscopy, and essential oil composition [J]. Journal of Forestry Research, 26 (2): 501-507.

XIE Y, HILL C A S, JALALUDIN Z, et al. 2011. The water vapor sorption behavior of three celluloses: Analysis using parallel exponential kinetics and interpretation using the Kelvin-Voigt viscoelastic model [J]. Cellulose, 18 (3): 517-530.

ZAIHAN J, HILL C A S, CURLING S, et al. 2009. Moisture adsorption isotherms of acacia mangium and endospermum malaccense using dynamic vapor sorption [J]. Journal of Tropical Forest Science, 21 (3): 277-285.

ZHANG J, DATTA A K. 2004. Some considerations in modeling of moisture transport in heating of hygroscopic materials [J]. Drying Technology, 22 (8): 1983-2008.

ZHANG X, KUNZEL H M, ZILLIG W, et al. 2016. A Fickian model for temperature-dependent sorption hysteresis in hygrothermal modeling of wood materials [J]. International Journal of Heat and Mass Transfer, 100: 58-64.

ZYLKA R, KLESZCZYNSKA H, KUPIEC J, et al. 2009. Modifications of erythrocyte membrane hydration induced by organic tin compounds [J]. Cell Biology International, 33 (7): 801-806.

Chapter 8 Molecular association of adsorbed water with heat-treated wood cell walls during moisture desorption process examined by micro-FTIR spectroscopy

8.1 Introduction

Wood is a natural lignocellulosic material (Alvarez et al. 2004; Garrote et al. 1999; Tong et al. 1990), which has been widely used in many aspects of our lives, such as construction, furniture manufacturing and interior decoration (Balat 2011; Fitzpatrick et al. 2010; Monlau et al. 2013; Sun and Cheng 2002). However, wood is inherently hygroscopic because it contains many hydrophilic groups (Esteban et al. 2006; Gauvin et al. 2014; Wang and Wang 1999; Zhang et al. 2016). When the wood is placed in a humid environment, most of its internal pores are occupied by water taken from the surrounding environment, and moisture content not only can seriously affect its properties, such as mechanical (Ates et al. 2009; Borrega and Karenlampi 2008; Korkut and Hiziroglu 2009; Missio et al. 2015; Moliński and Raczkowski 1988; Srinivas and Pandey 2012), elastic (Hogan Jr and Niklas 2004; Maeda and Fukada 1987; Ozyhar et al. 2013), and electrical properties (Khan et al. 1991), but also causes internal stresses in the wood, causing the wood to crack, warp and other defects, which seriously affects the use of wood and its products. In order to improve its hygroscopicity and expand its scope of application, it is necessary to modify the wood. In recent years, many methods have been advocated including plasticization, impregnation, acetylation, heat treatment, compression and bending, bleaching and dyeing, and other modification techniques, in which heat modification has been proven to significantly reduce moisture sorption and improve dimensional stability (Bakar et al. 2013; Borrega and Karenlampi 2010; Garcia et al. 2012; Gunduz et al. 2008; Scheiding et al. 2016). For example, Hill *et al.* believed that heat treatment technology can reduce the concentration of hygroscopic substances in wood components, reduce the hygroscopicity and internal stress of wood, and increase the dimensional stability of wood. Therefore, a fundamental understanding of moisture sorption in heat-treated wood is very important with respect to heat-treated wood's commercial utilization as well as its biology and chemistry.

Water sorption, the important property of heat-treated wood, has been studied in lots of literature (Ahmed et al. 2013; Dilik and Hiziroglu 2012; Metsä-Kortelainen and Viitanen 2012; Priadi and Hiziroglu 2013; Salca and Hiziroglu 2014; Wang et al. 2011; Willems et al. 2015). Kartal *et al.* (Kartal et al. 2007) studied water sorption of untreated and heat-treated Sugi wood, and determined that less water was absorbed by the heat-treated specimen. Besides, Metsä-Kortelainen *et al.* (Metsa-Kortelainen et al. 2006) examined the hygroscopic property of two kinds of wood, and found that heat treatment resulted in a decrease in the moisture content. Meanwhile, Hill *et al.* (Hill et al. 2012) studied moisture sorption properties of heat-treated scots pine using a dynamic vapor sorption (DVS) setup. Then, sorption isotherms and sorption hysteresis of other heat-treated wood including acacia, sesendok, norway spruce and eucalyptus pellita were all investigated using DVS method (Jalaludin et al. 2010; Sun et al. 2017; Willems 2014). Further, it was confirmed that the decrease in the hydroscopicity during heat treatment was due to the change in chemical composition of wood (Huang et al. 2012). Although the quantitatively analyzing water adsorption in heat-treated wood has been carried out, there has been scant research directed at molecular association between water and heat-treated wood during moisture desorption process, which is evidently the key perspective of moisture sorption mechanism.

As shown before, heat-treated wood contains many hydrophilic groups which are potential water sorption sites (Araujo et al. 1992; Gezici-Koc et al. 2017). Because of many possibilities of the formation of hydrogen bond at these sites and complex structures of adsorbed water deviated from ordinary water, molecular association between water and heat-treated wood are very confusing, which deserves a more detailed research. Meanwhile, many methods such as near-infrared (NIR) spectroscopy (Ferraz et al. 2005; Inagaki et al. 2008; Lestander 2008; Mora et al. 2011), nuclear magnetic resonance (NMR) spectroscopy (Bardet et al. 2004; Casieri et al. 2004; Hsi et al. 1977; Kekkonen et al. 2014; Senni et al. 2009), Fourier transform infrared (FTIR) spectroscopy (Almgren et al. 2008; Guo et al. 2017; Hakkou et al. 2005) and Raman spectroscopy (Atalla 1987; Ding et al. 2016) have been used to study the water sorption. For example, Berthold *et al.* and Tsuchikawa *et al.* (Berthold et al. 1998; Tsuchikawa and Siesler 2003) showed that NIR spectroscopy was sensitive enough to distinguish between water molecules adsorbed onto different hydrophilic groups. Zhang *et al.* (Zhang et al. 2013) used NMR technique to discriminate various structures of water based on spin-spin relaxation time, and showed that the water in yellow poplar (*Liriodendron tulipifera* L.) had five structures. Meanwhile, Agarwal et al. (Agarwal and Kawai 2005) demonstrated that water desorption in the cellulose filter paper and black spruce (*Picea mariana*) thermomechanical pulp has a linear dependence relation with the declining Raman intensity of the O—H stretching envelope. More importantly, Olsson et al. (Olsson and Salmen 2004) examined the sorption characteristics of some wood polymers by FTIR spectroscopy at nine levels of relative humidity ranging from 0% to 80%, confirmed the relation between water uptake and the increase of O—H stretching band, and indicated the chemical sites for water sorption. Furthermore, Célino *et al.* (Celino et al. 2014) showed FTIR characteristic information was available for qualitative analysis of moisture sorption mechanism. In these methods, FTIR spectroscopy owns many obvious advantages, i.e. the easily distinguishable characteristic peak (Abidi et al. 2014; Capron et al. 2007), the high sensitivity of water (Ping et al. 2001), and the good accuracy of spectral analysis (Moreira and Santos 2004). In recent years, micro-FTIR spectroscopy has been improved to own higher sensitivity in investigating molecular structure alteration (Igisu et al. 2012). Based on this, micro-FTIR spectroscopy has largest potential to characterize molecular association between water and heat-treated wood.

As micro-FTIR spectroscopy has the largest potential, the aim of this study is to apply this spectroscopic technique to demonstrate molecular association between water and a typical wood (i.e., heat-treated wood) during the moisture desorption process. Firstly, we measured heat-treated wood spectra in the RH region from high level to low level. Secondly, an analysis of these spectra and corresponding difference spectra was used to characterize effective water sorption sites and identify the main spectral region closely associated with moisture sorption. Finally, from component peak analysis of the identified spectral region, we demonstrated molecular structures of desorbed water during moisture desorption process.

8.2 Materials and methods

8.2.1 Materials

Wood specimens (dimensions 100 mm × 30 mm × 20 mm in length, width, and thickness) were cut from straight stem of Ginkgo biloba L. (Ginkgoaceae) wood. Then heat treatment was used for these wood specimens in electric vacuum drying oven under controlled condition of (180 ± 1) ℃. This heat treatment lasted 4 h. From these heat-treated wood specimens, transverse sections were prepared without embedding and any chemical treatment. These sections were cut using a manual rotary microtome (Leica RM2135), and then placed on the bottom of sample cell. Furthermore, this sample cell was filled with water-saturated N_2 at least three days before micro-FTIR spectral measurement.

8.2.2 Experimental instrument for spectral measurement

Diagrammatic sketch of experimental installation is presented in Figure 8.1(a). The primary section of the instrument was a spectrometer (Nicolet IN 10^{TM}), which was applied for recording micro-FTIR spectral development of heat-treated wood against RH during the RH region from 93.0% to 0% at the constant temperature of 25 ℃. This spectrometer included one microscope, which could be used for morphology observation. In spectral measurement, standard spectral acquisition was adopted, all micro-FTIR spectra were acquired from one randomly selected area (100 μm by 100 μm), and these spectra in the range of 720-4000 cm^{-1} were accumulated 32 times by scanning the grating at the spectral resolution of 4 cm^{-1}. Figure 8.1(b) also displays the diagrammatic sketch of sample cell. Firstly, one transverse section was mounted onto the bottom of the sample cell. Then, this cell was sealed, and placed on the automatic stage of micro-FTIR spectrometer. The RH was changed using saturated steam-air mixture which was accurately controlled by flow meter. More, the RH could be tested using humidity and temperature meter and real-time displayed on the computer.

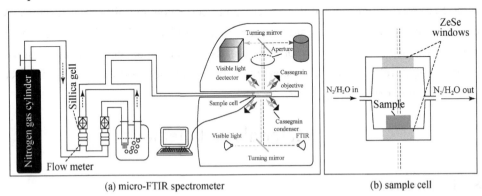

(a) micro-FTIR spectrometer　　　　　　　(b) sample cell

Figure 8.1　(a) Diagrammatic sketch of experimental installation; (b) Diagrammatic sketch of sample cell

Prior to recording micro-FTIR spectral development of heat-treated wood vs. RH, kinetic spectroscopy test was carried out to determine balance time. Once the target RH was set a new value, and a delay of about 5 min would occur. During this delay, actual RH became close to the target RH, and then remained almost unchanged afterwards. Meanwhile, after 15 min, the spectra recorded every one minute were kept constant. Based on the above results, 60 min were set to balance time for micro-FTIR spectral measurement during each RH level.

8.2.3 Micro-FTIR spectral data processing

1. Acquisition of Difference Spectra

To further qualitatively evaluating moisture sorption of heat-treated wood, difference spectrum approach should be applied, using which difference spectra were obtained via the original spectra subtracting the spectrum measured at 0% RH.

2. Application of Component Peak Analysis

Furthermore, component peak analysis was applied to distinguish between different types of bond water. In the process of implementation, the identified spectral region was fitted using the Pseudo-Voigt function in Origin 7.0 software. This function was linear combination of Gaussian function and Lorentzian function, as shown.

$$y = y_0 + A\left[m_u \cdot \frac{2}{\pi} \cdot \frac{w}{4(x-x_c)^2 + w^2} + (1-m_u) \cdot \frac{\sqrt{4\ln 2}}{\sqrt{\pi}w} \cdot e^{-\frac{4\ln 2}{w^2}(x-x_c)^2} \right] \quad (8.1)$$

In this equation, x was the independent variable, y was dependent variable, y_0 was baseline, A was peak area of the component peak, x_c was peak position of the component peak, w was full width at half maxima, and m_u was profile shape factor.

8.3 Results and discussion

8.3.1 Effective water sorption sites of heat-treated wood

Figure 8.2 shows heat-treated wood spectra recorded in the RH region from 93.0% to 0%. The development of micro-FTIR spectra with RH was shown clearly. With the decrease of RH, the peak situated at 3346 cm^{-1} from O—H stretching vibration was reduced, indicating that water was successively desorbed from the OH group. In the spectrum collected at 93.0% RH, the 1736 cm^{-1} band was derived from C=O stretching vibration, and the 1590 cm^{-1} band belonged to the aromatic skeletal plus the C=O stretching vibration. For comparison, the spectrum of heat-treated wood measured at 0% RH was also displayed, in which these two bands appeared at 1739 cm^{-1} and 1594 cm^{-1}. When the RH decreased, the band positions of these two bands exhibited blue shifts which indicated some water was desorbed from the C=O group.

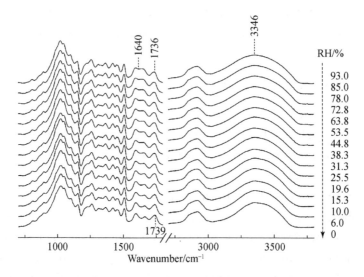

Figure 8.2 Micro-FTIR spectra collected during the RH region from 93.0% to 0%

In order to further extract precise spectral information about the moisture desorption process, difference spectrum was employed. Figure 8.3 presents difference spectra collected during RH decreased from 93.0% to 6.0%.

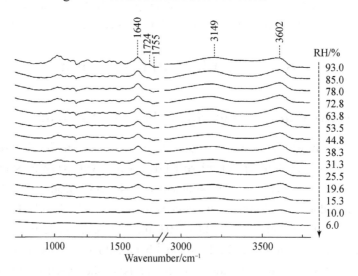

Figure 8.3 Difference spectra collected during RH decreased from 93.0% to 6.0%

The broad band situated between 3700 cm^{-1} and 2900 cm^{-1} was reduced *vs.* RH which was due to moisture desorption from the OH group of heat-treated wood. The same varying tendency has been found in the other lignocellulosic materials. Based on the above results, it was inferred that more accurate spectral region of this broad band correlated with moisture desorption was from 3700 cm^{-1} to 2900 cm^{-1}. Meanwhile, the 1755 cm^{-1} band belonged to free hydrogen bonded C=O group increased, while the 1724 cm^{-1} band was attributed to one hydrogen bonded C=O group of uronic acids in either xylan or pectin had reverse trend. Accordingly, the C=O group was confirmed to be closely associated with the moisture desorption process. To summarize, the results showed that the effective water sorption sites of heat-treated wood included OH and C=O groups and it provided foundation for demonstrating molecular structures of desorbed water in the moisture desorption process.

8.3.2　Molecular structure change of water

Previous studies have confirmed that the O—H stretching band of FTIR spectroscopy is broad, unresolved and generated from more than one type of water. Therefore, component peak analysis could be carried over into this broad band situated between 3700 cm^{-1} and 2900 cm^{-1} to distinguish different types of bound water, as described in the section of micro-FTIR spectral data processing. Here, three components were advocated. Among these components, two component peaks such as strongly and weakly bound waters were confirmed by Olsson *et al.* (Olsson and Salmen 2004) and the same parameters of these two types of bound water were used. Moreover, the third component was introduced to acquire more satisfactory fitting results, which was attributed to moderately bound water. It has been proved the presence of moderately bond water, so this introduction was reasonable. From component peak analysis, three component peaks were identified in difference spectra collected at various RH levels during the moisture desorption process. Difference spectra as squareness dot lines and the corresponding fitted peaks as red solid lines are shown in Figure 8.4, in which RH and coefficient of determination (R^2) are also demonstrated. In our fitting, these fitting bands lay at 3149 cm^{-1}, 3496 cm^{-1} and 3602 cm^{-1} separately, whose positions were same as those quoted in the literature (Guo et al. 2018). The frequency of FTIR spectral band is affected by hydrogen bonding, and some band can display a red shift along with the enhancement of hydrogen-bonded association. Therefore, the component peak in the higher frequency region (3149 cm^{-1}) was attributed to the strongly bond water, the component peak in the moderate frequency region (3496 cm^{-1}) was assigned to the moderately bond water, and the component peak in the lower frequency region (3602 cm^{-1}) belonged to the weakly bond water.

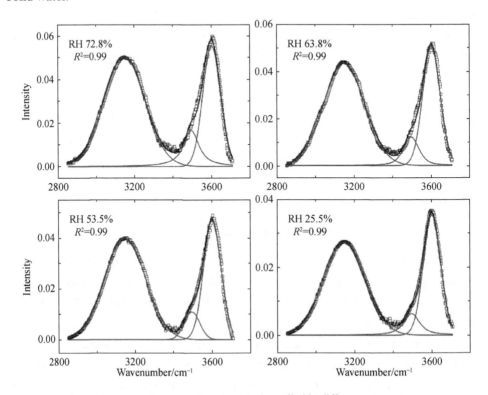

Figure 8.4　Component peak analysis applied in difference spectra

To illustrate how the bond water changed during the moisture desorption process, the integrated intensities of each component peak with RH are demonstrated in Figure 8.5. It was expected that the integrated intensities of these component peaks were reduced with the decrease of RH, for the water was desorbed during the moisture desorption process. Meanwhile, there were obviously different reducing trends for these component peaks. According to these reducing trends, the moisture desorption process could be separated into three parts (Part One, Part Two, and Part Three), as illustrated in Figure 8.5.

In Part One, the RH was higher than 47.0%. The integrated intensities of two component peaks (A_{3149} and A_{3602}) both decreased, and the reducing rate of A_{3149} was higher than that of A_{3602}. Meanwhile, the integrated intensity of the 3496 cm^{-1} band remained almost unchanged. These changes indicated that the desorbed strongly bond water was more than the desorbed weakly bond water in this part. Previous literatures found that under high RH condition, water molecules are connected together with strong hydrogen bonds, and form five-molecule tetrahedral structure (Guo et al. 2016). Based on that the desorption happened at higher RH level, and strongly bond water preferred to desorbed in this part, the desorbed water mainly owned the five-molecule tetrahedral structure, as illustrated in Figure 8.6 (a).

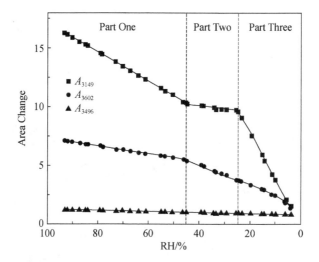

Figure 8.5 Variations of strongly bond water and weakly bond water with RH during the moisture desorption process, and k was the slope

In Part Two, the RH was in the range from 47.0% to 25.0%. A_{3149} and A_{3496} were almost constant, while A_{3602} was observed to decrease. It suggested that the main desorbed water in this part was weakly bond water. Therefore, in this part, the new desorbed water was those indirectly bond to the heat-treated wood through the existing water, and their structures should be HOH··· water, as illustrated in Figure 8.6 (b).

In Part Three, the RH was lower than 25.0%. All integrated intensities of three component peaks decreased, and the reducing rate of the 3149 cm^{-1} band was highest. These changes indicated that the strongly bond water was mostly desorbed in this part. As the moisture desorption happened at such lower RH, the desorbed water should be those bound directly to sorption sites. As mentioned before, the hydroxyl and carbonyl groups were effective water sorption sites of heat-treated wood. Therefore, in this part, molecular structures of desorbed water were $C=O\cdots(HOH)$ and $OH\cdots(OH_2)$, as illustrated in Figure 8.6 (c).

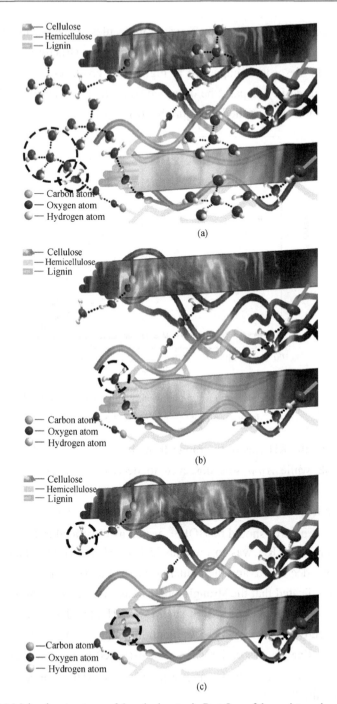

Figure 8.6 (a) Molecular structures of desorbed water in Part One of the moisture desorption process; (b) Molecular structures of desorbed water in Part Two of the moisture desorption process; (c) Molecular structures of desorbed water in Part Three of the moisture desorption process

8.4 Conclusions

Molecular structures of desorbed water were investigated using micro-FTIR spectroscopy during the moisture desorption process of a typical wood (i.e., heat-treated wood). A qualitative analysis of these spectra confirmed that $C=O$ and $O-H$ groups were effective water sorption sites and the spectral range of 3700-2900 cm^{-1} was closely associated with moisture desorption. Moreover, component peak analysis was carried over into this identified spectral range, in which three component peaks situated at 3149 cm^{-1}, 3496 cm^{-1} and 3602 cm^{-1} belonged to strongly, moderately and weakly bond water. There were obviously different reducing trends for these component peaks. According to these reducing trends, the moisture desorption process could be separated into three parts. In Part One, the desorbed water mainly owned the five-molecule tetrahedral structure; in Part Two, the desorbed water was those indirectly bond to the heat-treated wood through the existing water, and their structures should be $HOH\cdots$ water; in Part Three, the molecular structures of desorbed water were $C=O\cdots(HOH)$ and $OH\cdots(OH_2)$.

References

ABIDI N, CABRALES L, HAIGLER C H. 2014. Changes in the cell wall and cellulose content of developing cotton fibers investigated by FTIR spectroscopy [J]. Carbohydrate Polymers, 100: 9-16.

AGARWAL U P, KAWAI N. 2005. "Self-absorption" phenomenon in near-infrared Fourier transform Raman spectroscopy of cellulosic and lignocellulosic materials [J]. Applied Spectroscopy, 59 (3): 385-388.

AHMED S A, YANG Q, SEHLSTEDT-PERSSON M, et al. 2013. Accelerated mold test on dried pine sapwood boards: Impact of contact heat treatment [J]. Journal of Wood Chemistry and Technology, 33 (3): 174-187.

ALMGREN K M, AKERHOLM M, GAMSTEDT E K, et al. 2008. Effects of moisture on dynamic mechanical properties of wood fiber composites studied by dynamic FTIR spectroscopy [J]. Journal of Reinforced Plastics and Composites, 27 (16-17): 1709-1721.

ALVAREZ P, BLANCO C, SANTAMARIA R, et al. 2004. Improvement of the thermal stability of lignocellulosic materials by treatment with sulphuric acid and potassium hydroxide [J]. Journal of Analytical and Applied Pyrolysis, 72 (1): 131-139.

ARAUJO C D, MACKAY A L, HAILEY J R T, et al. 1992. Proton magnetic resonance techniques for characterization of water in wood: Application to white spruce [J]. Wood Science and Technology, 26 (2): 101-113.

ATALLA R H. 1987. Raman spectroscopy and the Raman microprobe: valuable new tools for characterizing wood and wood pulp fibers [J]. Journal of Wood Chemistry and Technology, 7 (1): 115-131.

ATES S, AKYILDIZ M H, OZDEMIR H. 2009. Effects of heat treatment on calabrian pine (pinus brutia ten.) wood [J]. BioResources, 4 (3): 1032-1043.

BAKAR B F A, HIZIROGLU S, TAHIR P M. 2013. Properties of some thermally modified wood species [J]. Materials and Design, 43: 348-355.

BALAT M. 2011. Production of bioethanol from lignocellulosic materials via the biochemical pathway: A review [J]. Energy Conversion and Management, 52 (2): 858-875.

BARDET M, FORAY M F, MARON S, et al. 2004. Characterization of wood components of Portuguese medieval dugout canoes with high-resolution solid-state NMR [J]. Carbohydrate Polymers, 57 (4): 419-424.

BERTHOLD J, OLSSON R, SALMEN L. 1998. Water sorption to hydroxyl and carboxylic acid groups in carboxymethylcellulose (CMC) studied with NIR-spectroscopy [J]. Cellulose, 5 (4): 281-298.

BORREGA M, KARENLAMPI P P. 2008. Mechanical behavior of heat-treated spruce (Picea abies) wood at constant moisture content and ambient humidity [J]. Holz als Roh und Werkstoff, 66 (1): 63-69.

BORREGA M, KARENLAMPI P P. 2010. Hygroscopicity of heat-treated Norway spruce (Picea abies) wood [J]. European Journal of Wood and Wood Products, 68 (2): 233-235.

CAPRON I, ROBERT P, COLONNA P, et al. 2007. Starch in rubbery and glassy states by FTIR spectroscopy [J]. Carbohydrate Polymers, 68 (2): 249-259.

CASIERI C, SENNI L, ROMAGNOLI M, et al. 2004. Determination of moisture fraction in wood by mobile NMR device [J]. Journal of Magnetic Resonance, 171 (2): 364-372.

CÉLINO A, GONCALVES O, JACQUEMIN F, et al. 2014. Qualitative and quantitative assessment of water sorption in natural fibers using ATR-FT'IR spectroscopy [J]. Carbohydrate Polymers, 101: 163-170.

DILIK T, HIZIROGLU S. 2012. Bonding strength of heat treated compressed eastern redcedar wood [J]. Materials and Design, 42: 317-320.

DING T, WANG C, PENG W. 2016. A theoretical study of moisture sorption behavior of heat-treated pine wood using Raman spectroscopic analysis [J]. Journal of Forestry Engineering, 1(5): 15-19.

ESTEBAN L G, FERNANDEZ F G, CASASUS A, G, et al. 2006. Comparison of the hygroscopic behavior of 205-year-old and recently cut juvenile wood from Pinus sylvestris L. [J]. Annals of Forest Science, 63 (3): 309-317.

FERRAZ A, MENDONCA R, GUERRA A, et al. 2005. Near - infrared spectra and chemical characteristics of pinus taeda (Loblolly pine) wood chips biotreated by the white-rot fungus ceriporiopsis subvermispora [J]. Journal of Wood Chemistry and Technology, 24 (2): 99-113.

FITZPATRICK M, CHAMPAGNE P, CUNNINGHAM M F, et al. 2010. A biorefinery processing perspective: treatment of lignocellulosic materials for the production of value-added products [J]. Bioresource Technology, 101 (23): 8915-8922.

GARCIA R A, de CARVALHO A M, de FIGUEIREDO LATORRACA J V, et al. 2012. Nondestructive evaluation of heat-treated Eucalyptus grandis Hill ex Maiden wood using stress wave method [J]. Wood Science and Technology, 46 (1-3): 41-52.

GARROTE G, DOMÍNGUEZ H, PARAJÓ J C. 1999. Hydrothermal processing of lignocellulosic materials [J]. Holz Als Roh- Und Werkstoff, 57 (3): 191-202.

GAUVIN C, JULLIEN D, DOUMALIN P, et al. 2014. Image correlation to evaluate the influence of hygrothermal loading on wood [J]. Strain, 50 (5SI): 428-435.

GEZICI-KOC O, ERICH S J F, HUININK H P, et al. 2017. Bound and free water distribution in wood during water uptake and drying as measured by 1D magnetic resonance imaging [J]. Cellulose, 24 (2): 535-553.

GUNDUZ G, NIEMZ P, AYDEMIR D. 2008. Changes in specific gravity and equilibrium moisture content in heat-treated fir (Abies nordmanniana subsp bornmulleriana Mattf.) wood [J]. Drying Technology, 26 (9): 1135-1139.

GUO X, QING Y, WU Y, et al. 2016. Molecular association of adsorbed water with lignocellulosic materials examined by micro-FTIR spectroscopy [J]. International Journal of Biological Macromolecules, 83: 117-125.

GUO X, WU Y, YAN N. 2017. Characterizing spatial distribution of the adsorbed water in wood cell wall of Ginkgo biloba L. by μ-FTIR and confocal Raman spectroscopy [J]. Holzforschung, 71 (5): 415-423.

GUO X, WU Y, YAN N. 2018. *In situ* micro-FTIR observation of molecular association of adsorbed water with heat-treated wood [J]. Wood Science and Technology, 52 (4): 971-985.

HAKKOU M, PETRISSANS M, ZOULALIAN A, et al. 2005. Investigation of wood wettability changes during heat treatment on the basis of chemical analysis [J]. Polymer Degradation and Stability, 89 (1): 1-5.

HILL C A S, RAMSAY J, KEATING B, et al. 2012. The water vapor sorption properties of thermally modified and densified wood [J]. Journal of Materials Science, 47 (7): 3191-3197.

HOGAN JR C J, NIKLAS K J. 2004. Temperature and water content effects on the viscoelastic behavior of Tilia americana(Tiliaceae) sapwood [J]. Trees, 18 (3): 339-345.

HSI E, HOSSFELD R, BRYANT R G. 1977. Nuclear magnetic resonance relaxation study of water absorbed on milled northern white-cedar [J]. Journal of Colloid and Interface Science, 62 (3): 389-395.

HUANG X, KOCAEFE D, KOCAEFE Y, et al. 2012. Changes in wettability of heat-treated wood due to artificial weathering [J]. Wood Science and Technology, 46 (6): 1215-1237.

IGISU M, TAKAI K, UENO Y, et al. 2012. Domain-level identification and quantification of relative prokaryotic cell abundance in microbial communities by Micro-FTIR spectroscopy [J]. Environmental Microbiology Reports, 4 (1): 42-49.

INAGAKI T, YONENOBU H, TSUCHIKAWA S. 2008. Near-infrared spectroscopic monitoring of the water adsorption/desorption process in modern and archaeological wood [J]. Applied Spectroscopy, 62 (8): 860-865.

JALALUDIN Z, HILL C A S, XIE Y, et al. 2010. Analysis of the water vapor sorption isotherms of thermally modified acacia and sesendok [J]. Wood Material Science and Engineering, 5: 194-203.

KARTAL S N, HWANG W, IMAMURA Y. 2007. Water absorption of boron-treated and heat-modified wood [J]. Journal of Wood Science, 53 (5): 454-457.

KEKKONEN P M, YLISASSI A, TELKKI V. 2014. Absorption of water in thermally modified pine wood as studied by nuclear magnetic resonance [J]. Journal of Physical Chemistry C, 118 (4): 2146-2153.

KHAN M A, ALI K M I, WANG W. 1991. Electrical properties and X-ray diffraction of wood and wood plastic composite (WPC) [J]. International Journal of Radiation Applications and Instrumentation. Part C. Radiation Physics and Chemistry, 38 (3): 303-306.

KORKUT S, HIZIROGLU S. 2009. Effect of heat treatment on mechanical properties of hazelnut wood (Corylus columa L.) [J]. Materials and Design, 30 (5): 1853-1858.

LESTANDER T A. 2008. Water absorption thermodynamics in single wood pellets modelled by multivariate near-infrared spectroscopy [J]. Holzforschung, 62 (4): 429-434.

MAEDA H, FUKADA E. 1987. Effect of bound water on piezoelectric, dielectric, and elastic properties of wood [J]. Journal of Applied Polymer Science, 33 (4): 1187-1198.

METSÄ-KORTELAINEN S, ANTIKAINEN T, VIITANIEMI P. 2006. The water absorption of sapwood and heartwood of Scots pine and Norway spruce heat-treated at 170 ℃, 190 ℃, 210 ℃ and 230 ℃ [J]. Holz Als Roh- Und Werkstoff, 64 (3): 192-197.

METSÄ-KORTELAINEN S, VIITANEN H. 2012. Wettability of sapwood and heartwood of thermally modified Norway spruce and Scots pine [J]. European Journal of Wood and Wood Products, 70 (1-3): 135-139.

MISSIO A L, MATTOS B D, de CADEMARTORI P H G, et al. 2015. Effects of two-step freezing-heat treatments on japanese raisintree (hovenia dulcis thunb.) wood properties [J]. Journal of Wood Chemistry and Technology, 36 (1): 16-26.

MOLIŃSKI W, RACZKOWSKI J. 1988. Mechanical stresses generated by water adsorption in wood and their determination by tension creep measurements [J]. Wood Science and Technology, 22 (3): 193-198.

MONLAU F, BARAKAT A, TRABLY E, et al. 2013. Lignocellulosic materials into biohydrogen and biomethane: Impact of structural features and pretreatment [J]. Critical Reviews in Environmental Science and Technology, 43 (3): 260-322.

MORA C R, SCHIMLECK L R, YOON S, et al. 2011. Determination of basic density and moisture content of loblolly pine wood disks using a near infrared hyperspectral imaging system [J]. Journal of Near Infrared Spectroscopy, 19 (5): 401-409.

MOREIRA J L, SANTOS L. 2004. Spectroscopic interferences in Fourier transform infrared wine analysis [J]. Analytica Chimica Acta, 513 (1): 263-268.

OLSSON A M, SALMEN L. 2004. The association of water to cellulose and hemicellulose in paper examined by FTIR spectroscopy [J]. Carbohydrate Research, 339 (4): 813-818.

OZYHAR T, HERING S, SANABRIA S J, et al. 2013. Determining moisture-dependent elastic characteristics of beech wood by means of ultrasonic waves [J]. Wood Science and Technology, 47 (2): 329-341.

PING Z H, NGUYEN Q T, CHEN S M, et al. 2001. States of water in different hydrophilic polymers-DSC and FTIR studies [J]. Polymer, 42 (20): 8461-8467.

PRIADI T, HIZIROGLU S. 2013. Characterization of heat treated wood species [J]. Materials and Design, 49: 575-582.

SALCA E, HIZIROGLU S. 2014. Evaluation of hardness and surface quality of different wood species as function of heat treatment [J]. Materials and Design, 62: 416-423.

SCHEIDING W, DIRESKE M, ZAUER M. 2016. Water absorption of untreated and thermally modified sapwood and heartwood of Pinus sylvestris L. [J]. European Journal of Wood and Wood Products, 74 (4): 585-589.

SENNI L, CASIERI C, BOVINO A, et al. 2009. A portable NMR sensor for moisture monitoring of wooden works of art, particularly of paintings on wood [J]. Wood Science and Technology, 43 (1-2): 167-180.

SRINIVAS K, PANDEY K K. 2012. Effect of heat treatment on color changes, dimensional stability, and mechanical properties of wood [J]. Journal of Wood Chemistry and Technology, 32 (4): 304-316.

SUN B, WANG Z, LIU J. 2017. Changes of chemical properties and the water vapor sorption of Eucalyptus pellita wood thermally modified in vacuum [J]. Journal of Wood Science, 63 (2): 133-139.

SUN Y, CHENG J Y. 2002. Hydrolysis of lignocellulosic materials for ethanol production: A review [J]. Bioresource Technology, 83 (1): 1-11.

TONG X, SMITH L H, MCCARTY P L. 1990. Methane fermentation of selected lignocellulosic materials [J]. Biomass, 21 (4): 239-255.

TSUCHIKAWA S, SIESLER H W. 2003. Near-infrared spectroscopic monitoring of the diffusion process of deuterium-labeled molecules in wood. Part I: Softwood [J]. Applied Spectroscopy, 57 (6): 667-674.

WANG S Y, WANG H L. 1999. Effects of moisture content and specific gravity on static bending properties and hardness of six wood species [J]. Journal of Wood Science, 45 (2): 127-133.

WANG S, MAHLBERG R, JAMSA S, et al. 2011. Surface properties and moisture behavior of pine and heat-treated spruce modified with alkoxysilanes by sol-gel process [J]. Progress in Organic Coatings, 71 (3): 274-282.

WILLEMS W. 2014. The water vapor sorption mechanism and its hysteresis in wood: The water/void mixture postulate [J]. Wood Science and Technology, 48 (3): 499-518.

WILLEMS W, ALTGEN M, MILITZ H. 2015. Comparison of EMC and durability of heat treated wood from high versus low water vapor pressure reactor systems [J]. International Wood Products Journal, 6 (1): 21-26.

ZHANG M, WANG X, GAZO R. 2013. Water states in yellow poplar during drying studied by time-domain nuclear magnetic resonance [J]. Wood and Fiber Science, 45 (4): 423-428.

ZHANG X, KUNZEL H M, ZILLIG W, et al. 2016. A Fickian model for temperature-dependent sorption hysteresis in hygrothermal modeling of wood materials [J]. International Journal of Heat and Mass Transfer, 100: 58-64.

Chapter 9 Molecular association of adsorbed water with cellulose during moisture adsorption process examined by micro-FTIR spectroscopy

9.1 Introduction

As a substitute for oil and other non- renewable energy, natural polymer has many advantages, such as large reserves, renewable, environmental protection, etc. In recent years its application has become a very active research field. Wood is one of the most abundant natural macromolecule raw materials in nature. It is well known that wood consists mainly of lignin, cellulose and hemicellulose. In different plants, the proportion of the three components of wood fiber is different. In general, lignin content is about 15%-30%, cellulose content is about 30%-50%, hemicellulose content is about 15%-25%. The cellulose in the cell wall is used as the skeleton material, which is filled into the matrix composed of a part of hemicellulose and lignin. The hemicellulose connects the cellulose and lignin. Cellulose is a kind of D-glucose chain macromolecule which is composed of beta-1,4 glycosidic bonds. Hemicellulose is a polysaccharide molecule composed of pentose and hexose. The main chain can be composed of one sugar group or two or more sugar groups containing hydroxyl, acetyl, carboxyl, methoxyl and the like groups. Compared with cellulose, cellulose is monosaccharide, and hemicellulose is not only a monosaccharide, but also an inhomogeneous monosaccharide. Different monosaccharides can be linked together to form different monosaccharides. Lignin is a complex amorphous polymer, which is made up of phenylpropane as a unit and linked by ether bond and carbon-carbon bond. Lignin has three basic structural units, i. e., syringyl, guaiacyl and *p*-hydroxyphenyl structures. Cellulose is embedded in hemicellulose and lignin as skeleton. Although cellulose has a simple basic chemical structure unit, the morphology and distribution of cellulose in biomass are very complex and anisotropic. The length of the natural cellulose molecular chain is about 5000 nm, equivalent to the length of the chain with 10000 glucose units. Due to the strong hydrogen bonding in and between the molecular chains of cellulose, the molecular chains of cellulose aggregate and form some crystallization regions, which make the crystallization region of cellulose have very high strength.

With the development of nano science and technology in various fields, the study of wood materials has changed from macroscopic to microcosmic and nano scale, and great breakthrough has been made. Therefore, the preparation of nano-fibers from lignocellulose is increasingly favored by researchers, and becomes a hot research field of cellulose. In recent years, many researchers have been able to produce nanoscale fiber products from different plant resources, which can be divided into two types: cellulose nanocrystalline whiskers and cellulose nanofibers. In order to make full use of the advantages of cellulose nanofibers, researchers are devoted to develop cellulose nanofibers into various functional materials and study their properties, which make cellulose nanofibers have great potential in the fields of flexible displays, solar cells, highly transparent membrane materials, adsorption filter materials, biomedicine and so on (Aytac et al. 2015; Balea et al. 2016; Koga et al. 2014; Rodriguez et al. 2012; Wang et al. 2018). Due to their outstanding properties such as high mechanical strength, and low thermal expansion coefficient, large elasticity, small diameter and so on (Gamelas et al. 2015; Ishii et al. 2011; Sun et al. 2015; Yue et al. 2012), cellulose nanofiber film prepared from woody raw materials is favorable for use in automotive components, tissue engineering and green packaging (Deng et al. 2017; Khalil et al. 2012; Khalil et al. 2016; Shi et al. 2014; Stelte and Sanadi 2009). There are a large number of free hydroxyl groups in cellulose molecules, which are easy to combine with water molecules to form hydrogen bonds, so cellulose has hygroscopic properties. The hygroscopicity increases with the increase of amorphous region, that is, with the decrease of crystallinity. In the process of desorption, the hydrogen bond is broken, which leads to the increase of free hydroxyl, and then adsorbs more water molecules, but in the process of desorption, the reformation of hydrogen bond is hindered by the internal structure. Therefore, cellulose will show hysteresis phenomenon. Cellulose is an insulator in the dry state. When it contains water, its conductivity will increase with the increase of water. So inherent hydrophilic groups of cellulose nanofiber film can cause water adsorption in hygrothermal conditions (Uraki et al. 2010), which has a large influence on surface behavior and leads to reliability problems like the degradation of dielectric properties, corrosion, or delamination (Isa et al. 2013; Vogt et al. 2005). Considering these impacts give rise to studying water adsorption mechanism of cellulose nanofiber film, which can assist in finding optimal strategies for reducing water adsorption.

Cellulose nanofiber film is a type of cellulose materials. For cellulose materials, water adsorption characteristics are prime properties that can be demonstrated using sigmoidal-shaped isotherms and analyzed using the theory of layered adsorption (Lequin et al. 2010). A number of experimental approaches have been used to characterize the water absorption of cellulose materials (Angkuratipakorn et al. 2017; Peresin et al. 2010; Wan et al. 2009; Watanabe et al. 2006), such as mass spectrometry (Zakharov et al. 2003), quartz crystal microgravimetry (Kittle et al. 2011), dielectric relaxation spectroscopy (Smith 1995; Sugimoto et al. 2008), differential scanning calorimetry (Agrawal et al. 2004; Mccrystal et al. 1997; Mccrystal et al. 1999; Nakamura et al. 1981; Nelson 1977), nuclear magnetic resonance spectroscopy (Bergenstrahle et al. 2008; Felby et al. 2008; Froix and Goedde 1976; Ogiwara et al. 1969; Ono et al. 1997; Topgaard and Soderman 2001), dynamic vapor sorption technique (Agrawal et al. 2004; Driemeier et al. 2012; Hill et al. 2010; Ono et al. 1997; Zaihan et al. 2009), and Fourier transform infrared (FTIR) spectroscopy (Murphy and Depinho 1995; Olsson and Salmen 2004). Of these, the latter method of FTIR spectroscopy has been proved to be well-adapted for investigating the water adsorption mechanism. This technique provides the advantages of being highly sensitive to water-detection (Ji et al. 1998; Kitano et al. 2005; Ludvigsson et al. 2000; Ping et al. 2001; Sammon et al. 1998), enabling an accurate quantitative analysis (Camacho and Karlsson 2001; Granada et al. 2012; Jamaludin et al. 2000; Namduri and Nasrazadani 2008; Reig et al. 2002), and being able to show structural information contained in the vibrational spectra at a molecular level (Faghihzadeh et al. 2016; Huy et al. 2004; Lefevre and Subirade 2000; Li et al. 2013; Tofan-Lazar and Al-Abadleh 2012). With developments of new technology, micro-FTIR spectroscopy has been recently introduced. It is also suitable for the quantitative analysis of nano- cellulose polymers and other compounds containing C—H, O—H, N—H groups. It has the advantages of simple preparation, high speed, no damage to sample composition and structure, and can be used for qualitative and quantitative analysis of sample composition and structure. In addition, as micro-FTIR spectroscopy equipped with an additional visible-light microscope can be used to visualize morphology and select the observation area of a micro-sized sample, it is thus considered to be a promising tool for studying water adsorption in cellulose nanofiber film.

Many promising results about the water adsorption mechanism of cellulose materials have been provided (Belbekhouche et al. 2011; Kachrimanis et al. 2006; Mihranyan et al. 2004; Okubayashi et al. 2005; Wolf et al. 1984). For example, Célino *et al.* (2014) showed that FTIR spectroscopy could be used for qualitatively and quantitatively analyzing the moisture absorption of natural fibers. Olsson et al.(2004) used FTIR spectroscopy to examine water adsorption and determine important relationship between moisture weight gain and increases in the O—H stretching envelope (Olsson and Salmen 2004). Kachrimanis et al.(2006) demonstrated that the chemical sites for water adsorption were OH and C=O groups for partly carboxymethylated cellulose (Kachrimanis et al. 2006). As differing chemical sites for water adsorption exist and there are different structural forms of adsorbed water that deviate from ordinary water, the structural changes in adsorbed water during the water adsorption process are highly complex. However, until recently there has been a lack of understanding about the molecular structure of adsorbed water, which is evidently an important part of the water adsorption mechanism.

The main purpose of this study was to characterize molecular structure of adsorbed water in cellulose nanofiber film. We used micro-FTIR spectrometer, which is an instrument that can obtain both the physical and chemical information of a sample and is valuable for the study of water adsorption, so we use it here and collect *in situ* spectra of a typical cellulose nanofiber film (i.e., cellulose nanofiber with cellulose I structure) over a wide range of RH levels, with the aid of a specially designed sample chamber. These micro-FTIR spectra with high resolution, high signal-to-noise ratio (SNR), high sensitive and real-time were applied to demonstrate chemical adsorption sites and molecular structure of adsorbed water.

9.2 Materials and methods

9.2.1 Materials

This work focused on a typical cellulose nanofiber film which was made from cellulose nanofiber with cellulose I structure. The preparation procedures for this film were clearly reported by Han *et al* (Han et al. 2013), therefore only a brief description was provided here. Dried bleached wood pulp (W-50 grade of KC Flock, Nippon Paper Chemicals Co., Tokyo, Japan) was used as the raw material; this was hydrolyzed for 1 h with 48 wt % sulfuric acid and then filtered under vacuum. The filtered material was mixed with distilled water thoroughly for 20 min and then centrifuged at 26 ℃ for three cycles. Cellulose nanofiber was obtained by centrifuging after each washing and dialyzed against distilled water in regenerated cellulose dialysis tubes with a molecular weight cut off of 12000-14000 (Fisher Scientific, Pittsburgh, PA, USA) for several days until a pH neutral solution was obtained. This suspension was then passed through a high-pressure homogenizer (Microfluidizer M-110P, Microfluidics Corp., Newton, MA, USA) at a rate of 135 mL/min for five passes. The concentration of homogenized cellulose nanofiber suspension was 0.7 wt %, and by adding water the concentration was adjusted to 0.1 wt %. The average length and width of the cellulose nanofiber obtained were estimated to be (732 ± 208)nm and (21 ± 7) nm, respectively, and the corresponding aspect ratio was approximately 35. The suspension obtained was then quickly frozen at −75 ℃ for about 2 h and freeze dried at a sublimating temperature of −88 ℃ under vacuum for three days to form a film. The final film with a thickness of about 20 μm was placed between two pieces of cover slips and then sealed in plastic bags. Prior to the experiment, the film was dried in a vacuum drying oven at (60 ± 3)℃ for 48 h.

9.2.2 Micro-FTIR spectroscopy setup

Figure 9.1 showed the experimental setup used to study water adsorption in cellulose nanofiber film. The apparatus consisted mainly of two functioning parts, a micro-FTIR system for measuring FTIR spectra and a specifically designed sample chamber for adjusting ambient RH levels. The central apparatus was the Nicolet IN 10^{TM} spectrometer (Thermo Electron Scientific Instruments, Madison, WI, USA), which was installed with an additional visible-light microscope in which two pathways were mounted for visible and IR beams (these could be alternated using the mirrors); the visible beam was from a visible light source and was used for visual examination and selecting the observation area. An aperture was placed on the optical path and the observation area (over which the IR spectrum was collected) was defined. The spatial resolution of the micro-FTIR spectroscope was limited to 6.25 μm by the wavelength and instrumental parameters, so the observation area of more than 6.25 μm by 6.25 μm was selected in this experiment. Once the microscope was set in IR transmission mode, we collected micro-FTIR spectra at 2 cm^{-1} spectral resolutions between 4000 cm^{-1} and 720 cm^{-1} and collected 16 scans. We successively measured the sample spectra and the corresponding background spectra at the selected RH levels to eliminate the influence of moisture. The process used to switch measurements between the sample and the corresponding background was listed as follows. Firstly, one observation area in the sample was randomly selected. The position of the chosen observation area was saved in a computer program and named the "sample point". Secondly, the observation area was adjusted to another position where there was no sample, and this position was also saved by the computer program and named the "background point". Thirdly, spectra were attained at the sample point, and then selected at the background point. All movements and spectral acquisitions were automatically accomplished by the computer program.

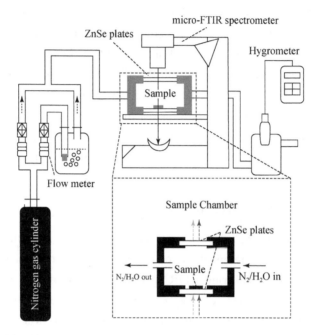

Figure 9.1　The schematic diagram of experimental setup used to study water adsorption in cellulose nanofiber film

Figure 9.1 also showed a schematic of the specially designed sample chamber. The sample was mounted onto a ZnSe plate fixed to the bottom of the chamber, which was then sealed by a lid composed of a ZnSe plate. The RH in this chamber was adjusted by mixing a stream of water-saturated N_2 and dry N_2 at controlled flow rates. After changing the RH to a new setting, there was typically a delay of 3-5 min before it stabilized. A kinetic spectroscopy test was conducted to calculate the time taken to reach the desired RH, in which the sample was equilibrated at the starting RH, i.e., 0%RH, for 6 h. We took the first spectrum measurement 10 s after starting, after which we changed the RH of the sample chamber to 96% and then recorded the spectra every 1 min. No changes were detected in the measured spectra after a period of 20 min, and this situation remained the same after 6 h. Based on these kinetic spectroscopy test results, we maintained a conditioning time of 60 min at each value prior to measurement to allow the sample to stabilize at the specific RH. The resulting RH was then measured using a hygrometer (Center 310, Center Technology Corp., New Taipei, Taiwan, China), which was installed in the outlet of the sample chamber.

9.2.3 Data processing

Difference spectra were introduced by FTIR subtractive spectroscopy technique to clearly characterize the minor structural changes. Using this technique, difference spectrum at the specific RH was acquired via subtracting the spectrum of sample at 0% RH from that at the specific RH at ratio 1 ∶ 1.

In further analysis of the difference spectra obtained, component band analysis was used to identify more than one component. Component band analysis was performed using one mixing formula that contained the Gaussian and Lorentzian formula shown blow as,

$$y = y_0 + A\left[m_u \cdot \frac{2}{\pi} \cdot \frac{w}{4(x-x_c)^2 + w^2} + (1-m_u) \cdot \frac{\sqrt{4\ln 2}}{\sqrt{\pi} w} \cdot e^{-4\ln 2 / w^2 \cdot (x-x_c)^2} \right] \quad (9.1)$$

In this formula, y_0 was the baseline offset of the curve, A was the peak area, x_c was the peak position, w was the full widths at half-maximum of the fitting peak, and m_u was the profile shape factor. In addition, A and m_u of the three components (i.e., strongly, moderately, and weakly hydrogen bonded water) were varied in the component band analysis, while x_c and w of the three components were kept unchanged.

9.3 Results and discussion

9.3.1 Micro-FTIR spectra of cellulose nanofiber film

To investigate the water adsorbed by cellulose nanofiber film, we randomly selected three observation areas (A, B, and C) which were 50 μm × 50 μm in the cellulose nanofiber film. Figure 9.2(a) showed a visible light microscopy image of the cellulose nanofiber film, illustrating the position, size, and surface morphology of the three randomly selected observation areas. Micro-FTIR spectra from the observation area A at different RH levels (from 0% to 90%) were displayed in Figure 9.2(b), where it could be seen that the most intense peak at 3352 cm^{-1} increased with a rise in RH; this was attributed to the O—H stretching band of cellulose nanofiber film and absorbed water. In addition, an increase was observed in the weaker peak at around 1643 cm^{-1} in all spectra, corresponding to the O—H bending of adsorbed water (Wandlowski et al. 2004). The spectral changes of these two peaks indicated that water molecules were successively absorbed by OH groups of cellulose nanofiber film. The peaks at 1204 cm^{-1} and 1036 cm^{-1} at 0% RH belonged to C—OH in plane bending at C-6, and C—O stretching at C-6 (Oh et al. 2005). When the RH was increased from 0% to 90%, these two peaks shifted to 1201 cm^{-1} and 1033 cm^{-1}, respectively. The results indicated that the O(6)—H groups of cellulose nanofiber film formed hydrogen bonds with the absorbed water. The results of this study showed that the OH groups were the main adsorption sites, and this was considered to be valuable knowledge in the characterization of molecular structure of adsorbed water.

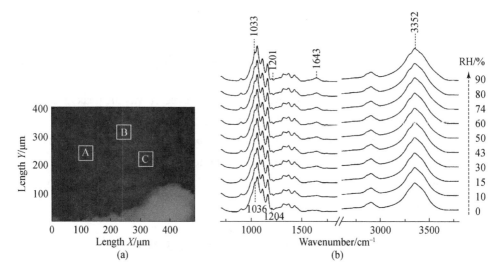

Figure 9.2 (a) Visible light microscopy image of the cellulose nanofiber film. The red hollow boxes (A, B, and C) indicated the positions of three randomly selected observation areas. (b) Micro-FTIR spectra from observation area A during the water adsorption process. The arrow showed the value of RH was from 0 to 90%

9.3.2 Difference spectra of cellulose nanofiber film at various RH levels

Water adsorption in micro-FTIR spectroscopy provided only an unresolved broad band, which offered limited precise information. This study further aimed to extract information about minor structural changes occurring with respect to water adsorption, and thus difference spectra were employed.

Figure 9.3 showed difference spectra of the cellulose nanofiber film in the RH range from 10% to 90%. The signature of the broad band in the spectral region of 2900-3700 cm^{-1} increased with the rise of RH; this was identified and assigned to absorbed water. Similar growth trends of this broad band have been evident in the measurements of other cellulosic materials. So it suggested that this cellulose nanofiber film had great water absorption capacity. In addition, the positive-going peak at 1646 cm^{-1} changed continuously, which further confirmed this conclusion. The peak centered at 1172 cm^{-1} was observed to decrease in intensity, while the peak at 1159 cm^{-1} was observed to rise. The changes of these two peaks were related to the whole chain of cellulose affected by the adsorbed water.

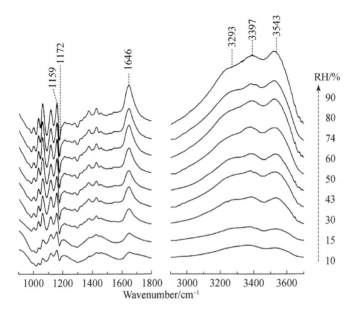

Figure 9.3 Difference micro-FTIR spectra of cellulose nanofiber film during the water adsorption process. The arrow showed the value of RH was from 10 to 90%

9.3.3 Different types of water adsorbed by cellulose nanofiber film

It has been shown that the band shape can indicate the presence of more than one component; therefore, the broad band in the spectral region of 2900-3700 cm^{-1} (as shown in Figure 9.3) was considered to arise from many types of adsorbed water. Component band analysis was thus applied to this broad band to extract the component peaks, as described in the 9.2.3 section. Three component peaks were determined in our work. As two components (strongly and weakly hydrogen bonded water) were observed by Olsson et al.(2004) to preliminarily show the water bonded to the O—H groups of cellulose and hemicelluloses, we used the same peak positions and FWHM here. However, the third component peak, which was denoted as moderately hydrogen bonded water, also needed to be included in the fitting to obtain satisfactory results for the cellulose nanofiber film. As existence of moderately hydrogen bonded water was identified, it was thus reasonable to introduce this component (Dong et al. 2007). For the third component peak, the peak position and its FWHM were fixed in our fitting process, as for the other component peaks. Results of component band analysis were shown in Figure 9.4; these included all three component peaks and correlation coefficients (R^2). As seen from Figure 9.4, the three component peaks at 3293 cm^{-1}, 3397 cm^{-1}, and 3543 cm^{-1} were assigned to strongly, moderately, and weakly hydrogen bonded water.

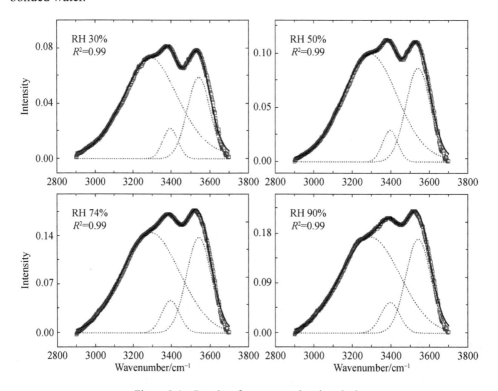

Figure 9.4　Results of component band analysis

These three types of adsorbed water were proved, and their structures were proposed to be "···HOH···", "····HOH···", and "····HOH····". Meanwhile, the areas of three component peaks at different RH levels were reported in Figure 9.5. As expected, the signatures of these three types of hydrogen bonded water were observed to rise with an increase in RH, for cellulose nanofiber film is easy to adsorb water by hydrogen bond. However, the growth rates of these three component peaks were significantly different. Based on the growth regularities, the water adsorption process of cellulose nanofiber film was divided into three stages (as shown in Figure 9.5).

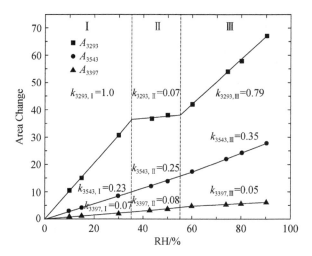

Figure 9.5 The areas of three component peaks ascribed to strongly, moderately and weakly hydrogen bonded water as a function of RH with a y-axis error bar derived from the variation between three replicates. k was the slope of area change, and two subscripts of k indicated the component peak and the stage

In the first stage, RH was lower than 35%. Although the areas of three component peaks at 3293 cm^{-1}, 3397 cm^{-1}, and 3543 cm^{-1} (A_{3293}, A_{3397}, and A_{3543}) increased obviously, the growth rate of A_{3293} was highest. This growth rate indicated that strongly hydrogen bonded water were mostly adsorbed during this stage. Considering that the strong hydrogen bonds were formed at the lower RH levels, the absorbed water should be bounded directly to the chemical adsorption sites of cellulose nanofiber film. As before, the hydroxyl groups were certified as main adsorption sites, which were demonstrated in Figure 9.6 (a). In addition, the structure of the water molecules forming strong hydrogen bonds was proposed as "···HOH···". Thus, in this stage, most of water absorbed by cellulose nanofiber film was CNF···HOH···CNF, as illustrated in Figure 9.6 (b).

In the second stage, RH was in the range from 35% to 55%. The peak areas of 3293 cm^{-1} and 3397 cm^{-1} maintained an almost original value, while the peak area at 3543 cm^{-1} showed an obvious increase. This indicated that most of the water absorbed in this stage was weakly hydrogen bonded water. This growth trend implied that the first layer of cellulose nanofiber film was nearly filled with adsorbed water as the RH was lower than 35%. Therefore, in this stage, the new absorbed water indirectly bonded to the cellulose nanofiber film through the existing absorbed water. The structure of these water molecules forming weak hydrogen bonds was proposed as "····HOH····", and thus most of water absorbed by cellulose nanofiber film in this stage was water····HOH····water [as shown in Figure 9.6 (c)].

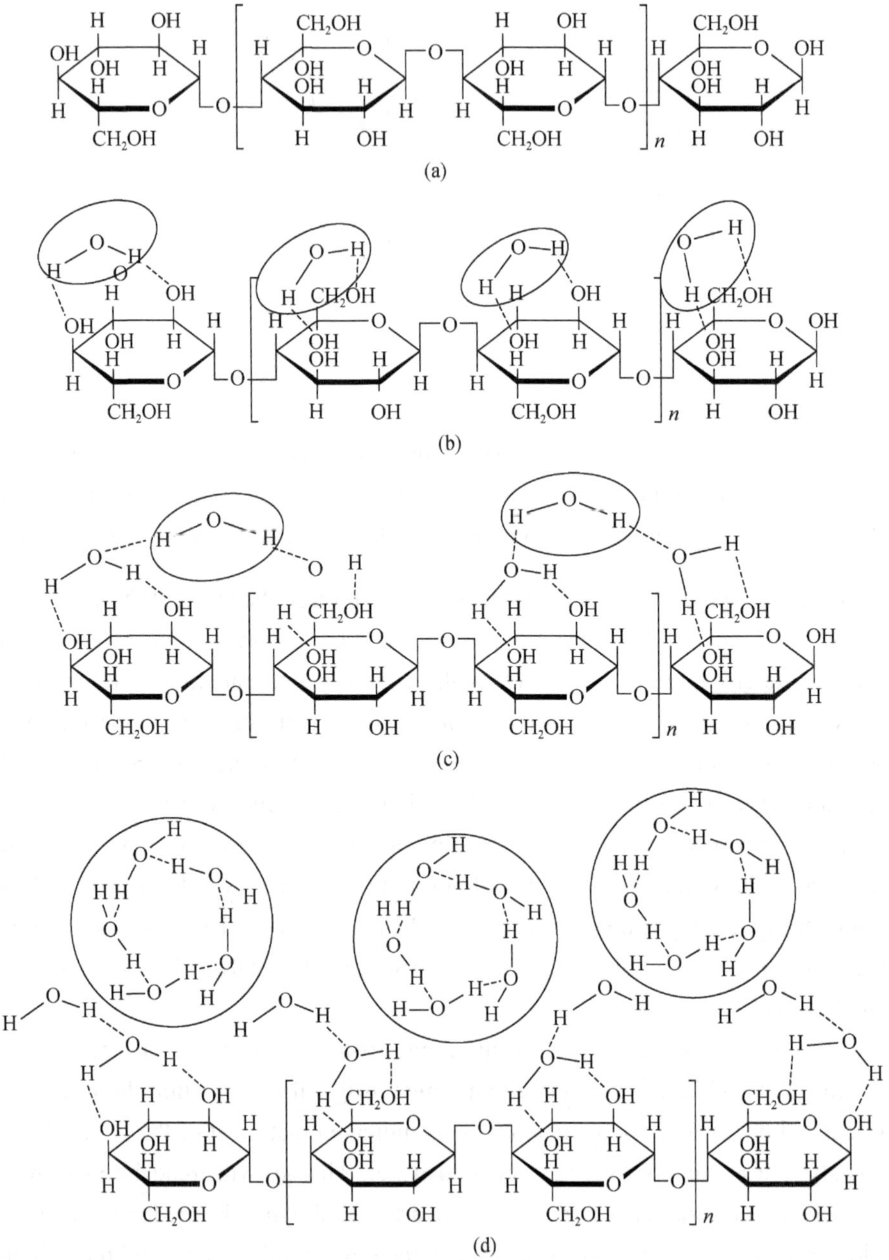

Figure 9.6 Proposed molecular structures of water mostly adsorbed in three stages

(a) The main adsorption sites of cellulose nanofiber film; (b) As RH was lower than 35%, most of water absorbed by cellulose nanofiber film was CNF···HOH···CNF; (c) As RH increased from 35% to 55%, most of absorbed water was water····HOH····water; (d) As RH was higher than 55%, most of absorbed water owned five-molecule tetrahedral structure

In the third stage, RH was higher than 55%. The areas of two peaks at 3293 cm^{-1} and 3543 cm^{-1} (A_{3293} and A_{3543}) increased continuously, while the growth rate of A_{3293} was higher than that of A_{3543}. Meanwhile, the original value of the peak area at 3397 cm^{-1} was almost maintained. All this suggested that the formed strongly hydrogen-bonded water was more than the formed weakly hydrogen bonded water in this stage. Considering that the hydrogen bond in the fully hydrogen-bonded five-molecule tetrahedral structure is known to be strong hydrogen bond, during this stage most of the water absorbed by cellulose nanofiber film would thus have five-molecule tetrahedral structure, as shown in Figure 9.6 (d).

9.4 Conclusions

The molecular structure of adsorbed water in cellulose nanofiber film was characterized using *in situ* micro-FTIR spectroscopy. The high SNR of micro-FTIR spectra of a typical cellulose nanofiber film (i.e., cellulose nanofiber with cellulose I structure) at various relative humidity enabled demonstration of the chemical adsorption sites for adsorbed water. In addition, component band analysis of micro-FTIR spectra showed that the water adsorption process of cellulose nanofiber film could be divided into three stages, and most of the water absorbed by cellulose nanofiber film in these three stages was CNF⋯HOH⋯CNF, water⋯⋯HOH⋯⋯water, and five-molecule tetrahedral structure, respectively. This study would provide a valuable insight into the water adsorption mechanism of cellulose nanofiber film, which could assist in finding optimal strategies for reducing water adsorption.

References

AGRAWAL A M, MANEK R V, KOLLING W M, et al. 2004. Water distribution studies within microcrystalline cellulose and chitosan using differential scanning calorimetry and dynamic vapor sorption analysis [J]. Journal of Pharmaceutical Sciences, 93 (7): 1766-1779.

ANGKURATIPAKORN T, SINGKHONRAT J, CHRISTY A A. 2017. Comparison of water adsorption properties of cellulose and cellulose nanocrystals studied by near-infrared spectroscopy and gravimetry [J]. Key Engineering Materials, 735: 235-239.

AYTAC Z, SEN H S, DURGUN E, et al. 2015. Sulfisoxazole/cyclodextrin inclusion complex incorporated in electrospun hydroxypropyl cellulose nanofibers as drug delivery system [J]. Colloids and Surfaces B-Biointerfaces, 128: 331-338.

BALEA A, MERAYO N, FUENTE E, et al. 2016. Valorization of corn stalk by the production of cellulose nanofibers to improve recycled paper properties [J]. BioResources, 11 (2): 3416-3431.

BELBEKHOUCHE S, BRAS J, SIQUEIRA G, et al. 2011. Water sorption behavior and gas barrier properties of cellulose whiskers and microfibrils films [J]. Carbohydrate Polymers, 83 (4): 1740-1748.

BERGENSTRAHLE M, WOHLERT J, LARSSON P T, et al. 2008. Dynamics of cellulose-water interfaces: NMR spin-lattice relaxation times calculated from atomistic computer simulations [J]. Journal of Physical Chemistry B, 112 (9): 2590-2595.

CAMACHO W, KARLSSON S. 2001. NIR, DSC, and FTIR as quantitative methods for compositional analysis of blends of polymers obtained from recycled mixed plastic waste [J]. Polymer Engineering and Science, 41 (9): 1626-1635.

CÉLINO A, GONCALVES O, JACQUEMIN F, et al. 2014. Qualitative and quantitative assessment of water sorption in natural fibers using ATR-FTIR spectroscopy [J]. Carbohydrate Polymers, 101: 163-170.

DENG Z, JUNG J, ZHAO Y. 2017. Development, characterization, and validation of chitosan adsorbed cellulose nanofiber (CNF) films as water resistant and antibacterial food contact packaging [J]. Lwt-Food Science and Technology, 83: 132-140.

DONG J, LI X, ZHAO L, et al. 2007. Raman observation of the interactions between NH_4^+, SO_4^{2-}, and H_2O in supersaturated $(NH_4)_2SO_4$ droplets [J]. Journal of Physical Chemistry B, 111 (42): 12170-12176.

DRIEMEIER C, MENDES F M, OLIVEIRA M M. 2012. Dynamic vapor sorption and thermoporometry to probe water in celluloses [J]. Cellulose, 19 (4): 1051-1063.

FAGHIHZADEH F, ANAYA N M, SCHIFMAN L A, et al. 2016. Fourier transform infrared spectroscopy to assess molecular-level changes in microorganisms exposed to nanoparticles [J]. Nanotechnology for Environmental Engineering, 1 (1): 1.

FELBY C, THYGESEN L G, KRISTENSEN J B, et al. 2008. Cellulose-water interactions during enzymatic hydrolysis as studied by time domain NMR [J]. Cellulose, 15 (5): 703-710.

FROIX M F, GOEDDE A O. 1976. The effect of temperature on the cellulose/water interaction from NMR relaxation times [J]. Macromolecules, 9 (3): 428-430.

GAMELAS J A F, PEDROSA J, LOURENCO A F, et al. 2015. On the morphology of cellulose nanofibrils obtained by TEMPO-mediated oxidation and mechanical treatment [J]. Micron, 72: 28-33.

GRANADA E, EGUIA P, VILAN J A, et al. 2012. FTIR quantitative analysis technique for gases. application in a biomass thermochemical process [J]. Renewable Energy, 41 (1): 416-421.

HAN J, ZHOU C, WU Y, et al. 2013. Self-assembling behavior of cellulose nanoparticles during freeze-drying: effect of suspension concentration, particle size, crystal structure, and surface charge [J]. Biomacromolecules, 14 (5): 1529-1540.

HILL C A S, NORTON A J, NEWMAN G. 2010. The water vapor sorption properties of Sitka spruce determined using a dynamic vapor sorption apparatus [J]. Wood Science and Technology, 44 (3): 497-514.

HUY T A, ADHIKARI R, LUPKE T, et al. 2004. Molecular deformation mechanisms of isotactic polypropylene in α-and β-crystal forms by FTIR spectroscopy [J]. Journal of Polymer Science Part B, 42 (24): 4478-4488.

ISA A, MINAMINO J, MIZUNO H, et al. 2013. Increased water resistance of bamboo flour/ polyethylene composites [J]. Journal of Wood Chemistry and Technology, 33 (3): 208-216.

ISHII D, SAITO T, ISOGAI A. 2011. Viscoelastic evaluation of average length of cellulose nanofibers prepared by TEMPO-mediated oxidation [J]. Biomacromolecules, 12 (3): 548-550.

JAMALUDIN S, AZLAN M, FUAD M, et al. 2000. Quantitative analysis on the grafting of an aromatic group on polypropylene in melt by FTIR technique [J]. Polymer Testing, 19 (6): 635-642.

JI S F, JIANG T L, XU K, et al. 1998. FTIR study of the adsorption of water on ultradispersed diamond powder surface [J]. Applied Surface Science, 133 (4): 231-238.

KACHRIMANIS K, NOISTERNIG M F, GRIESSER U J, et al. 2006. Dynamic moisture sorption and desorption of standard and silicified microcrystalline cellulose [J]. European Journal of Pharmaceutics and Biopharmaceutics, 64 (3): 307-315.

KHALIL H P S A, BHAT A H, YUSRA A F I. 2012. Green composites from sustainable cellulose nanofibrils: a review [J]. Carbohydrate Polymers, 87 (2): 963-979.

KHALIL H P S A, DAVOUDPOUR Y, SAURABH C K, et al. 2016. A review on nanocellulosic fibers as new material for sustainable packaging: process and applications [J]. Renewable and Sustainable Energy Reviews, 64: 823-836.

KITANO H, MORI T, TAKEUCHI Y, et al. 2005. Structure of water incorporated in sulfobetaine polymer films as studied by ATR-FTIR [J]. Macromolecular Bioscience, 5 (4): 314-321.

KITTLE J D, DU X, JIANG F, et al. 2011. Equilibrium water contents of cellulose films determined via solvent exchange and quartz crystal microbalance with dissipation monitoring [J]. Biomacromolecules, 12 (8): 2881-2887.

KOGA H, NOGI M, KOMODA N, et al. 2014. Uniformly connected conductive networks on cellulose nanofiber paper for transparent paper electronics [J]. NPG Asia Materials, 6 (3): e93.

LEFEVRE T, SUBIRADE M. 2000. Interaction of beta-lactoglobulin with phospholipid bilayers: a molecular level elucidation as revealed by infrared spectroscopy [J]. International Journal of Biological Macromolecules, 28 (1): 59-67.

LEQUIN S, CHASSAGNE D, KARBOWIAK T, et al. 2010. Adsorption equilibria of water vapor on cork [J]. Journal of Agricultural and Food Chemistry, 58 (6): 3438-3445.

LI W, PIERRELOUIS A, KWON K D, et al. 2013. Molecular level investigations of phosphate sorption on corundum (α-Al_2O_3) by ^{31}P solid state NMR, ATR-FTIR and quantum chemical calculation [J]. Geochimica Et Cosmochimica Acta, 107 (Complete): 252-266.

LUDVIGSSON M, LINDGREN J, TEGENFELDT J. 2000. FTIR study of water in cast Nafion films [J]. Electrochimica Acta, 45 (14): 2267-2271.

MCCRYSTAL C B, FORD J L, RAJABISIAHBOOMI A R. 1997. A study on the interaction of water and cellulose ethers using differential scanning calorimetry [J]. Thermochimica Acta, 294 (1): 91-98.

MCCRYSTAL C B, FORD J L, RAJABI-SIAHBOOMI A R. 1999. Water distribution studies within cellulose ethers using differential scanning calorimetry. 1. effect of polymer molecular weight and drug addition [J]. Journal of Pharmaceutical Sciences, 88 (8): 792-796.

MIHRANYAN A, LLAGOSTERA A P, KARMHAG R, et al. 2004. Moisture sorption by cellulose powders of varying crystallinity [J]. International Journal of Pharmaceutics, 269 (2): 433-442.

MURPHY D, DEPINHO M N. 1995. An ATR-FTIR study of water in cellulose-acetate membranes prepared by phase inversion [J]. Journal of Membrane Science, 106 (3): 245-257.

NAKAMURA K, HATAKEYAMA T, HATAKEYAMA H. 1981. Studies on bound water of cellulose by differential scanning calorimetry [J]. Textile Research Journal, 51 (9): 607-613.

NAMDURI H, NASRAZADANI S. 2008. Quantitative analysis of iron oxides using Fourier transform infrared spectrophotometry [J]. Corrosion Science, 50 (9): 2493-2497.

NELSON R A. 1977. The determination of moisture transitions in cellulosic materials using differential scanning calorimetry [J]. Journal of Applied Polymer Science, 21 (3): 645-654.

OGIWARA Y, KUBOTA H, HAYASHI S, et al. 1969. Studies of water adsorbed on cellulosic materials by a high resolution NMR spectrometer [J]. Journal of Applied Polymer Science, 13 (8): 1689.

OH S Y, YOO D I, SHIN Y, et al. 2005. FTIR analysis of cellulose treated with sodium hydroxide and carbon dioxide [J]. Carbohydrate Research, 340 (3): 417-428.

OKUBAYASHI S, GRIESSER U J, BECHTOLD T. 2005. Moisture sorption/desorption behavior of various manmade cellulosic fibers [J]. Journal of Applied Polymer Science, 97 (4): 1621-1625.

OLSSON A M, SALMEN L. 2004. The association of water to cellulose and hemicellulose in paper examined by FTIR spectroscopy [J]. Carbohydrate Research, 339 (4): 813-818.

ONO H, INAMOTO M, OKAJIMA K, et al. 1997. Spin-lattice relaxation behavior of water in cellulose materials in relation to the tablet forming ability of microcrystalline cellulose particles [J]. Cellulose, 4 (2): 57-73.

PERESIN M S, HABIBI Y, VESTERINEN A, et al. 2010. Effect of moisture on electrospun nanofiber composites of poly(vinyl alcohol) and cellulose nanocrystals [J]. Biomacromolecules, 11 (9): 2471-2477.

PING Z H, NGUYEN Q T, CHEN S M, et al. 2001. States of water in different hydrophilic polymers -DSC and FTIR studies [J]. Polymer, 42 (20): 8461-8467.

REIG F B, ADELANTADO J, MORENO M. 2002. FTIR quantitative analysis of calcium carbonate (calcite) and silica (quartz) mixtures using the constant ratio method. application to geological samples [J]. Talanta, 58 (PII S0039-9140(02)00372-74): 811-821.

RODRIGUEZ K, GATENHOLM P, RENNECKAR S. 2012. Electrospinning cellulosic nanofibers for biomedical applications: Structure and *in vitro* biocompatibility [J]. Cellulose, 19 (5): 1583-1598.

SAMMON C, MURA C, YARWOOD J, et al. 1998. FTIR-ATR studies of the structure and dynamics of water molecules in polymeric matrixes. a comparison of PET and PVC [J]. Journal of Physical Chemistry B, 102 (18): 3402-3411.

SHI X, ZHENG Y, WANG G, et al. 2014. pH and electro-responsive characteristics of bacterial cellulose nanofiber sodium alginate hybrid hydrogels for the dual controlled drug delivery [J]. RSC Advances, 4 (87): 47056-47065.

SMITH G. 1995. Dielectric analysis of water in microcrystalline cellulose [J]. Pharmacy and Pharmacology Communications, 1 (9): 419-422.

STELTE W, SANADI A R. 2009. Preparation and characterization of cellulose nanofibers from two commercial hardwood and softwood pulps [J]. Industrial and Engineering Chemistry Research, 48 (24): 11211-11219.

SUGIMOTO H, MIKI T, KANAYAMA K, et al. 2008. Dielectric relaxation of water adsorbed on cellulose [J]. Journal of non-Crystalline Solids, 354 (27): 3220-3224.

SUN X, WU Q, REN S, et al. 2015. Comparison of highly transparent all-cellulose nanopaper prepared using sulfuric acid and TEMPO-mediated oxidation methods [J]. Cellulose, 22 (2): 1123-1133.

TOFAN-LAZAR J, AL-ABADLEH H A. 2012. Kinetic ATR-FTIR studies on phosphate adsorption on iron (oxyhydr) oxides in the absence and presence of surface arsenic: molecular-level insights into the ligand exchange mechanism [J]. Journal of Physical Chemistry A, 116 (41): 10143-10149.

TOPGAARD D, SODERMAN O. 2001. Diffusion of water absorbed in cellulose fibers studied with ^{1}H-NMR [J]. Langmuir, 17 (9): 2694-2702.

URAKI Y, MATSUMOTO C, HIRAI T, et al. 2010. Mechanical effect of acetic acid lignin adsorption on honeycomb-patterned cellulosic films [J]. Journal of Wood Chemistry and Technology, 30 (4): 348-359.

VOGT B D, SOLES C L, LEE H J, et al. 2005. Moisture absorption into ultrathin hydrophilic polymer films on different substrate surfaces [J]. Polymer, 46 (5): 1635-1642.

WAN Y Z, LUO H, HE F, et al. 2009. Mechanical, moisture absorption, and biodegradation behaviors of bacterial cellulose fibre-reinforced starch biocomposites [J]. Composites Science and Technology, 69 (7): 1212-1217.

WANDLOWSKI T, ATAKA K, PRONKIN S, et al. 2004. Surface enhanced infrared spectroscopy - Au(1 1 1-20 nm)/sulphuric acid-new aspects and challenges [J]. Electrochimica Acta, 49 (8):

1233-1247.

WANG Y, CHENG Z, LIU Z, et al. 2018. Cellulose nanofibers/polyurethane shape memory composites with fast water-responsivity [J]. Journal of Materials Chemistry B, 6 (11): 1668-1677.

WATANABE A, MORITA S, OZAKI Y. 2006. A study on water adsorption onto microcrystalline cellulose by near-infrared spectroscopy with two-dimensional correlation spectroscopy and principal component analysis [J]. Applied Spectroscopy, 60 (9): 1054-1061.

WOLF W, SPIESS W E L, JUNG G, et al. 1984. The water-vapor sorption isotherms of microcrystalline cellulose (MCC) and of purified potato starch. results of a collaborative study [J]. Journal of Food Engineering, 3 (1): 51-73.

YUE Y, ZHOU C, FRENCH A D, et al. 2012. Comparative properties of cellulose nano-crystals from native and mercerized cotton fibers [J]. Cellulose, 19 (4): 1173-1187.

ZAIHAN J, HILL C A S, CURLING S, et al. 2009. Moisture adsorption isotherms of acacia mangium and endospermum malaccense using dynamic vapor sorption [J]. Journal of Tropical Forest Science, 21 (3): 277-285.

ZAKHAROV A G, PELIPETS O V, VORONOVA M, et al. 2003. Application of mass spectrometry for water sorption study on cellulose materials [J]. Journal of Molecular Liquids, 103: 161-167.

Chapter 10 Spatial distribution of adsorbed water in cellulose film studied using micro-FTIR spectroscopy

10.1 Introduction

Cellulose nanofiber film, which is a novel derivation of lignocellulosic material, has excellent physical, chemical, and biological properties (Ang-Atikarnkul et al. 2014; Zhang et al. 2017), such as low thermal expansion coefficient, high mechanical strength, reproducibility and environmental friendly (Balea et al. 2016; Ishii et al. 2011; Isogai et al. 2011; Liu et al. 2016; Yue et al. 2012), and it is very promising candidate for tissue engineering, electronics, and green packaging materials (Deng et al. 2017; Ghaderi et al. 2014; Manhas et al. 2015; Shi et al. 2014). However, hydrophilic cellulose nanofiber film adsorbs water under hydrothermal conditions (Newman and Davidson 2004; Uraki et al. 2010), which strongly affects surface behavior and leads to reliability problems (Isa et al. 2013; Vogt et al. 2005). Consequently, water adsorption of cellulose nanofiber film needs to be fully understood in order for this sustainable raw material to be efficiently utilized.

Water adsorption is a key property of lignocellulosic material (Bratasz et al. 2012; Khali and Rawat 2000; Nakano 1999; Ohmae et al. 2009; Tshabalala et al. 1999), and many experimental approaches have been used to characterize the water distribution (Hartley and Schneider 1993; Hernandez and Caceres 2010; Hunter 1995; Mao et al. 2014; Pearce et al. 1997; Taniguchi et al. 1978). The nuclear magnetic resonance (NMR) spectroscopy has been the most widely used and has provided vast amounts of water distribution data (Brownstein 1980; Casieri et al. 2004; Li et al. 1992; Menon et al. 1987; Rosenkilde and Glover 2002). Rosenkilde *et al.* (Rosenkilde and Glover 2002) used this technique to identify spatially resolved water adsorption in the lignocellulosic material at a spatial resolution of 20 mm. Another non-destructive technique for studying the water distribution is the computed tomographic (CT) scanner based on X-rays or gamma rays (Davis et al. 1993; Krueger et al. 2011; Sakata et al. 2009;

Toney et al. 1995; Zhou et al. 2000). By this approach, Fromm et al.(2001) investigated the water distribution in the lignocellulosic material at a spatial resolution of 0.1225 mm^3. Neutron radiography is also suitable for investigating the distribution of water (Hendrickx et al. 2017; Islam et al. 2003; Rosner et al. 2012). Mannes et al. (Mannes et al. 2009) confirmed that the absorbed water was quantified by this technique at a spatial resolution of 100 μm. Although the results are promising, each of the above techniques suffers from the resolution-related issues. More recently, micro-Fourier Transform Infrared (FTIR) technique has been improved and equipped with an additional microscope that offers the possibility for visual selecting observation areas (Liu et al. 2008). These techniques yield high spatial resolutions at the μm scale, and could provide observations with high-resolution of water *in-situ*, which may complement the results obtained by other techniques.

FTIR spectroscopy shows some distinct advantages because its spectral information allows both qualitative and quantitative analyses of the water adsorption (Belton et al. 1995). For example, Célino et al. (Célino et al. 2014) confirmed that the FTIR spectroscopy had the potential to qualitatively analyze water absorption of natural fibers. Olsson et al. (Olsson and Salmen 2004) used FTIR spectroscopy to examine water adsorption of cellulose materials, and determined that there was an important relationship between the weight gain due to moisture and increases in its O—H stretching envelope. Haxaire et al. (Haxaire et al. 2003) used this method to determine whether OH groups were bonded with water. Hofstetter et al. (Hofstetter et al. 2006) examined the molecular association of adsorbed water with cellulose via FTIR spectroscopy and showed the role that different hydrogen bonds in water adsorption. Kachrimanis et al. (Kachrimanis et al. 2006) and Berthold et al. (Berthold et al. 1998) both demonstrated that the chemical sites for water adsorption for partly carboxymethylated cellulose were OH and C=O groups. Guo et al. (Guo et al. 2017) confirmed that carboxyl C=O and C—O groups as well as OH groups were active sites for water adsorption in wood specimen. These works given as examples in this paragraph are very important, as they have provided much FTIR spectral information for qualitatively characterizing water adsorption of cellulose materials (Célino et al. 2014; Guo et al. 2018; Olsson and Salmen 2004). However, little attention has been devoted to investigate water distribution in cellulose nanofiber film using FTIR spectroscopy.

As micro-FTIR spectroscopy seems to be one promising experimental tool for studying water distribution (Hao et al. 2007; Ito and Nakashima 2002), the goal of the present study is to investigate the possibility of this approach. Firstly, we collected *in situ* micro-FTIR spectra of cellulose nanofiber film over a wide range of RH levels to identify spectral peaks associated with water adsorption. Secondly, based on these identified spectral peaks, we used micro-FTIR imaging approach to demonstrate the spatial distribution of cellulose and OH group in a randomly selected mapping area in cellulose nanofiber film at four different RH levels of 0%, 30%, 60% and 90%. Finally, difference 2-D micro-FTIR images of OH group distribution were introduced to visualize water adsorption of cellulose nanofiber film.

10.2 Materials and methods

10.2.1 Materials

This work analyzes a typical specimen of nanocellulose material (i.e., cellulose nanofiber film). The preparation for this cellulose nanofiber film is described clearly in earlier reports (Han et al. 2013). Therefore only a brief description of the preparation is provided here. Dried bleached wood pulp (W-50 grade from KC Flock, Nippon Paper Chemicals Co., Tokyo, Japan) was used as the raw material; this was hydrolyzed for 1 h with 48 wt % sulfuric acid and then filtered under vacuum. The filtered material was mixed with distilled water thoroughly for 20 min and then centrifuged at 25 ℃ for three cycles. The suspension was obtained by centrifuging after each washing and dialyzed against distilled water in regenerated cellulose dialysis tubes with a molecular weight cutoff of 12000-14000 (Fisher Scientific, Pittsburgh, PA, USA). This process was repeated for several days until a pH neutral solution was obtained. This material was then passed through a high-pressure homogenizer (Microfluidizer M-110P, Microfluidics Corp., Newton, MA, USA) at a rate of 135 mL/min for five times. After homogenization, the concentration of cellulose nanofiber suspension was 1.0 wt % and was adjusted to 0.1 wt % by adding water. The obtained suspensions of this nanocellulose sample were then quickly frozen at −75 ℃ for about 2 h and freeze-dried at a sublimating temperature of −88 ℃ under vacuum for three days to form a film. The final film was placed between two cover slips and then was sealed in plastic bags.

10.2.2 Micro-FTIR experimental setup

The instrument is presented in Figure 10.1. The main part of the apparatus was the micro-FTIR system (Nicolet IN 10^{TM} micro-FTIR spectrometer; Thermo Electron Scientific Instruments, Madison, WI, USA) This micro-FTIR spectrometer was equipped with an additional microscope, which was for the visual selection of the observation area. The visible and IR pathways could be alternately changed by turning the mirrors. In this experiment, two types of spectroscopic measurements were introduced such as standard spectral acquisition (spectral collection at the same point) and imaging acquisition (spectral collection at varying sample points). Standard spectral acquisition was used for characterizing the development of cellulose nanofiber film *vs*. RH in the water adsorption process, in which the spectra at different RH levels were collected from one randomly selected area. And the micro-FTIR spectra at 4 cm^{-1} spectral resolutions between 4000 cm^{-1} and 720 cm^{-1} were obtained with the grating operating 32 scans. Meanwhile, imaging acquisition was chosen for generating micro-FTIR images to determine cellulose and adsorbed water distributions. Here, one mapping area of 160 by 120 μm^2 was randomly selected in cellulose nanofiber film and the observation points of 20 by 20 μm^2 were mainly distributed in the selected mapping area. Spectra were obtained sequentially from a series of these observation points using point-by-point scanning. The spectral and spatial information was all recorded by a 2-D CCD detector.

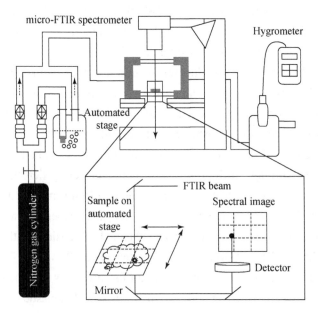

Figure 10.1   Sketch of the experimental apparatus for the micro-FTIR measurements

Figure 10.1 also shows a schematic of the specially designed sample chamber. Once the cellulose nanofiber film was placed on the bottom of the chamber (which was composed of a ZnSe plate), the sample chamber was sealed by a lid (which contained another ZnSe plate). The ZnSe plates provided the optical paths for both the visible and

IR light through this sample chamber. The RH in this chamber was adjusted by mixing a stream of dry nitrogen and saturated vapor, whose ratio was changed using a high-precision gas micro-flow meter (Alicat Scientific, Tucson, AZ, USA). The resulting RH was then measured using a humidity/temperature meter (Center 310, Center Technology Corp., New Taipei, Taiwan, China).

In the experiment, even if the RH of the chamber reached the set value, a period of 60 min should be maintained. The equilibrium time was from a kinetic spectroscopic test. Before the start of this test, cellulose nanofiber film was equilibrated at starting RH of 0% for 4 h. We took the first spectrum measurement after starting this process, after which we changed the RH of the sample chamber to 5% and then recorded the spectra every 0.5 min. As shown in Figure 10.2, after changing the target RH to a new setting, i.e., 5%RH, there was typically a delay of approximately 3-4 min in which the actual RH approached the set target RH. While, there were no changed detected in the measured spectra after a period of 15 min, and this situation remained the same after 100 min (the peak height of the major peak at 3348 cm^{-1} was used to demonstrate the spectral change). And then the target RH was changed to next setting, i.e., 10%RH, no change was detected in the measured spectra after a period of 15 min. In addition, the temperature in the specially designed sample chamber was maintained at 25 ℃. Based on these kinetic spectroscopy test results, we maintained the equilibrium time of 60 min.

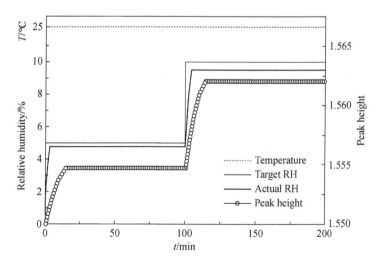

Figure 10.2 Typical changes of target RH, actual RH and peak height of the major peak at 3348 cm^{-1} during the water adsorption process. The temperature is maintained at 25 ℃

10.3 Results and discussion

10.3.1 Qualitatively analyzing water adsorption in cellulose nanofiber film

Figure 10.3 shows the visible light image of cellulose nanofiber film with a randomly selected mapping area (160 μm by 120 μm) and the positions of each observation point (63 points). In order to clearly demonstrate minor spectral changes of cellulose nanofiber film during water adsorption process, spectra from the whole mapping area (red box shown in Figure 10.3) were obtained over a wide range of RH levels from 0% to 90% (Figure 10.4). As shown in Figure 10.4, the most intense peak at 3348 cm^{-1} was attributed to O—H stretching vibration of cellulose nanofiber film and absorbed water (Zuluaga et al. 2009), and it was also observed to increase as the RH increased from 0 to 90%. The spectral change of this peak confirmed that OH groups were effective adsorption sites for adsorbed water (Guo et al. 2016). In addition, an increase was observed in the weaker peak at around 1635 cm^{-1}, corresponding to the O—H bending of adsorbed water. Moreover, the peak at 1160 cm^{-1} was assigned to that of C—O—C asymmetric stretching vibration at the β-glucosidic linkage, which was related to the cellulose chain (Kondo 1997). A continuous shift of this peak to lower wave numbers was observed as the RH increased. This shift was also caused by water adsorption, for absorbed water could stiffen the cellulose chain. Moreover, the broad peak of 2905 cm^{-1} was attributed to CH and CH$_2$ stretching vibrations (Gwon et al.

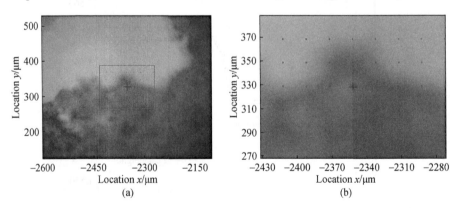

Figure 10.3 Visible light image of the cellulose nanofiber film. The red box indicates the position of the randomly selected mapping area including observation points (160 μm by 120 μm at 20 μm resolution, 63 observation points). In imaging acquisition mode, spectra are obtained sequentially from a series of observation points using point-by-point scanning

2010), and the peak at 1371 cm^{-1} was assigned to CH bending vibrations (Oh et al. 2005). In addition, the peaks at 1428 cm^{-1} and 1317 cm^{-1} were assigned to symmetric CH$_2$ bending and wagging vibrations of CH$_2$OH at C-6, respectively (El-Wakil and Hassan 2008; Oh et al. 2005). For these four peaks, there appeared to be little difference in the peak height measured at various RH levels, implying that these peaks rarely affected by water adsorption. Based on these characteristics, the most intense peak at 3348 cm^{-1} was found to be significantly impacted by water adsorption, while the four peaks at 2905 cm^{-1}, 1428 cm^{-1}, 1371 cm^{-1} and 1317 cm^{-1} were nearly not changed during the water adsorption process.

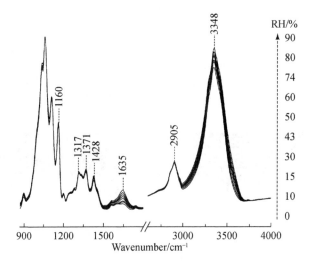

Figure 10.4 Micro-FTIR spectra at various RH levels as indicated

10.3.2 Spatial distribution of cellulose in the cellulose nanofiber film

Micro-FTIR imaging has the potential to provide useful information regarding spatial distribution of some chemical components (Kammer et al. 2010; Wehbe et al. 2013). Hence, by using spectral peaks assigned to cellulose, micro-FTIR image could be generated to determine the spatial distribution of cellulose in cellulose nanofiber film.

In the micro-FTIR spectrum of cellulose nanofiber film at the RH of 0%, the intense contributions from CH and CH$_2$ vibrations of cellulose were seen at 2905 cm^{-1}, 1428 cm^{-1}, 1371 cm^{-1} and 1317 cm^{-1} (Figure 10.4). Here, these four peaks were all used to generate 2-D micro-FTIR images of cellulose distribution for comparison [Figure 10.5(a)-(d)]. As shown in Figure 10.5(a), high cellulose concentration points (red) were occupied the most parts, and some locations where the cellulose concentration was low (blue) also could be observed. Cellulose concentration in this mapping area was nonuniform and varied significantly. There was no obvious differentiation among these four micro-FTIR images. Moreover, contour plots of these micro-FTIR images of cellulose distribution were similar to those of the visible light image (Figure 10.3).

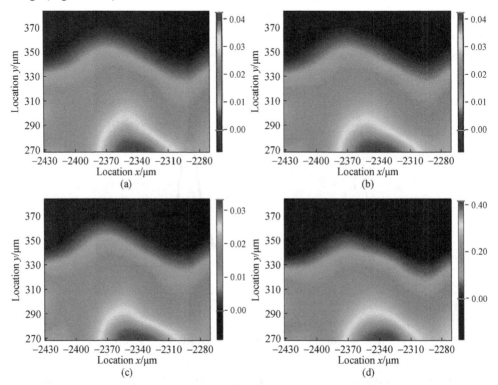

Figure 10.5 2-D micro-FTIR images (false color) of cellulose spatial distribution. Intensity scale appears on the right. Red locations indicate high concentration of cellulose; blue regions indicate low concentration

(a) This 2-D micro-FTIR image is generated using band of 2905 cm^{-1}; (b) This 2-D micro-FTIR image is generated using band of 1371 cm^{-1}; (c) This 2-D micro-FTIR image is generated using band of 1428 cm^{-1}; (d) This 2-D micro-FTIR image is generated using band of 1317 cm^{-1}

10.3.3 Spatial distribution of adsorbed water in the cellulose nanofiber film

As shown in Figure 10.4, both peaks at 1635 cm^{-1} and 3348 cm^{-1} were confirmed to be closely associated with water absorption. Compared with the peak at 3348 cm^{-1}, the peak at 1635 cm^{-1} was much weaker which may be more likely to be affected by the signal noise; therefore, the peak at 3348 cm^{-1} was used to generate 2-D micro-FTIR image. In order to visualize the water adsorption, micro-FTIR imaging was employed to investigate the same mapping area of cellulose nanofiber film at four different RH levels of 0%, 30%, 60%, and 90%. 2-D micro-FTIR images of OH group distribution at these four RH levels were all shown in Figure 10.6(a)-(d). In Figure 10.6(a), high OH group concentration spots (red spots) were found at the bottom central of the selected mapping area. Meanwhile, there existed many spots (blue spots) where the concentration of OH group was low, such as lower left corner and lower right corner of the selected mapping area. Therefore, OH group concentration in the selected mapping area was nonuniform. Furthermore, Figure 10.6(a)-(d) demonstrated the development of OH group distribution as the RH increased from 0% to 90%. However, the OH group distribution showed no significant variations with a rise in RH, especially in the case of comparing Figure 10.6(a) (RH 0%) with Figure 10.6(b) (RH 30%). This unclear variability could be explained that the signature of OH group from cellulose nanofiber film may hide the change of that from adsorbed water affected by water adsorption.

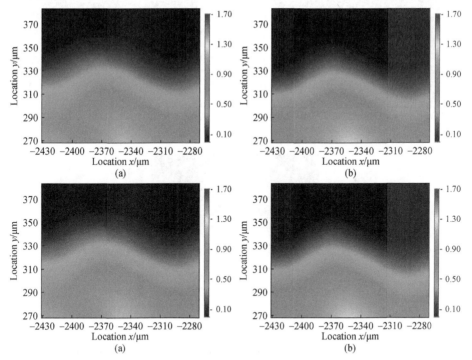

Figure 10.6   2-D micro-FTIR images (false color) of OH group distribution. Intensity scale appears on the right. Red locations indicate high concentration of OH group; blue regions indicate low concentration

(a) This 2-D micro-FTIR image is acquired at RH 0; (b) This 2-D micro-FTIR image is acquired at RH 30%; (c) This 2-D micro-FTIR image is acquired at RH 60%; (d) This 2-D micro-FTIR image is acquired at RH 90%

With the purpose of extracting information about adsorbed water, FTIR subtractive spectroscopy technique was introduced. It has been shown that using this technique, the difference spectrum at the specific RH can be acquired via subtracting the spectrum of sample at 0% RH from that at the specific RH, and it can be used to quantify the amount of water adsorbed by the sample (Schuttlefield et al. 2007). Here, using the same technique, difference 2-D micro-FTIR images were generated via subtracting the 2-D micro-FTIR image measured at 0% RH from those measured at different RH levels. And then, difference 2-D micro-FTIR images could also be used to determine the distribution of adsorbed water. Figure 10.7 shows these difference 2-D micro-FTIR images at three different RH levels of 30%, 60% and 90%. It appeared that the adsorbed water in the selected mapping area increased with a higher RH. Meanwhile, the concentration of adsorbed water on the bottom center of the selected

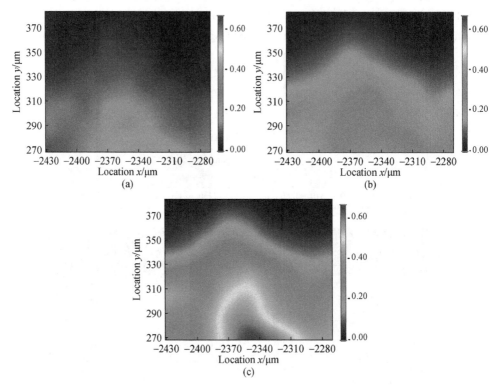

Figure 10.7   Difference 2-D micro-FTIR images (false color) characterizing adsorbed water distribution. Intensity scale appears on the right. Red locations indicate high concentration of adsorbed water; blue regions indicate low concentration

(a) This difference 2-D micro-FTIR image characterizes the water adsorbed below RH 30%; (b) This difference 2-D micro-FTIR image characterizes the water adsorbed below RH 60%; (c) This difference 2-D micro-FTIR image characterizes the water adsorbed below RH 90%

mapping area was higher than that in other regions. This distribution may be due to the higher cellulose concentration on the bottom center of the selected mapping area which was demonstrated in Figure 10.5(a). Moreover, the amount of adsorbed water in different regions of the selected mapping area was nonuniform, while the spots of high adsorbed water concentration spots in the selected mapping area (red spots) were not changed. More importantly, the results showed that the micro-FTIR imaging can provide useful information about the spatial distribution of adsorbed water at a high spatial resolution of 20 μm. Future objectives are improving spatial resolution of this technique to demonstrate adsorbed water distribution in detail by applying more advanced sample chamber, automated piezoelectric x, y mapping stage, and so on.

10.4 Conclusions

In situ visualization of water adsorption in cellulose nanofiber film with spatial micrometer resolution was achieved using micro-FTIR imaging. As the RH increased from 0% to 90%, the increase of the characteristic peaks at 3348 cm^{-1} indicated that OH groups were effective adsorption sites for adsorbed water. Meanwhile, the spectral peaks at 2905 cm^{-1}, 1428 cm^{-1}, 1371 cm^{-1} and 1317 cm^{-1} were assigned to the contribution of cellulose. On this basis, four 2-D micro-FTIR images of cellulose distribution were generated using these four peaks, and indicated that cellulose concentration in the selected mapping area was nonuniform. Meanwhile, using the most intense peak of 3348 cm^{-1}, four 2-D micro-FTIR image of OH group distribution were generated at four different RH levels of 0%, 30%, 60%, and 90%. With the purpose of extracting information about adsorbed water from these 2-D micro-FTIR images of OH group distribution, FTIR subtractive spectroscopy technique was used to generate difference 2-D micro-FTIR images which determined the distribution of adsorbed water clearly. Moreover, these difference 2-D micro-FTIR images at three different RH levels of 30%, 60% and 90% demonstrated the development of spatial distribution of adsorbed water in cellulose nanofiber film during water adsorption process. This study confirmed that the effectiveness of micro-FTIR imaging in visualizing water adsorption of cellulose nanofiber film, which could assist in developing a non-destructive and rapid method for *in situ* visualization of heterogeneous reaction with high spatial resolution of 20 μm.

References

ANG-ATIKARNKUL P, WATTHANAPHANIT A, RUJIRAVANIT R. 2014. Fabrication of cellulose nanofiber/chitin whisker/silk sericin bionanocomposite sponges and characterizations of their physical and biological properties [J]. Composites Science and Technology, 96: 88-96.

BALEA A, MERAYO N, FUENTE E, et al. 2016. Valorization of corn stalk by the production of cellulose nanofibers to improve recycled paper properties [J]. BioResources, 11 (2): 3416-3431.

BELTON P S, COLQUHOUN I J, GRANT A, et al. 1995. FTIR and NMR studies on the hydration of a high-M(r) subunit of glutenin. [J]. International Journal of Biological Macromolecules, 17 (2): 74-80.

BERTHOLD J, OLSSON R, SALMEN L. 1998. Water sorption to hydroxyl and carboxylic acid groups in carboxymethylcellulose (CMC) studied with NIR-spectroscopy [J]. Cellulose, 5 (4): 281-298.

BRATASZ L, KOZLOWSKA A, KOZLOWSKI R. 2012. Analysis of water adsorption by wood using the Guggenheim-Anderson-de Boer equation [J]. European Journal of Wood and Wood Products, 70 (4): 445-451.

BROWNSTEIN K R. 1980. Diffusion as an explanation of observed NMR behavior of water absorbed on wood [J]. Journal of Magnetic Resonance, 40 (3): 505-510.

CASIERI C, SENNI L, ROMAGNOLI M, et al. 2004. Determination of moisture fraction in wood by mobile NMR device [J]. Journal of Magnetic Resonance, 171 (2): 364-372.

CÉLINO A, GONCALVES O, JACQUEMIN F, et al. 2014. Qualitative and quantitative assessment of water sorption in natural fibers using ATR-FTIR spectroscopy [J]. Carbohydrate Polymers, 101: 163-170.

DAVIS J R, ILIC J, WELLS P. 1993. Moisture content in drying wood using direct scanning Gamma-ray densitometry [J]. Wood and Fiber Science, 25 (2): 153-162.

DENG Z, JUNG J, ZHAO Y. 2017. Development, characterization, and validation of chitosan adsorbed cellulose nanofiber (CNF) films as water resistant and antibacterial food contact packaging [J]. Lwt-Food Science and Technology, 83: 132-140.

EL-WAKIL N A, HASSAN M L. 2008. Structural changes of regenerated cellulose dissolved in FeTNa, NaOH/thiourea, and NMMO systems [J]. Journal of Applied Polymer Science, 109 (5): 2862-2871.

FROMM J H, SAUTTER I, MATTHIES D, et al. 2001. Xylem water content and wood density in spruce and oak trees detected by high-resolution computed tomography [J]. Plant Physiology, 127 (2): 416-425.

GHADERI M, MOUSAVI M, YOUSEFI H, et al. 2014. All-cellulose nanocomposite film made from bagasse cellulose nanofibers for food packaging application [J]. Carbohydrate Polymers, 104: 59-65.

GUO X, LIU L, WU J, et al. 2018. Qualitatively and quantitatively characterizing water adsorption of a cellulose nanofiber film using micro-FTIR spectroscopy [J]. RSC Advances, 8 (8): 4214-4220.

GUO X, QING Y, WU Y, et al. 2016. Molecular association of adsorbed water with lignocellulosic materials examined by micro-FTIR spectroscopy [J]. International Journal of Biological Macromolecules, 83: 117-125.

GUO X, WU Y, YAN N. 2017. Characterizing spatial distribution of the adsorbed water in wood cell wall of Ginkgo biloba L. by μ-FTIR and confocal Raman spectroscopy [J]. Holzforschung, 71 (5): 415-423.

GWON J G, LEE S Y, DOH G H, et al. 2010. Characterization of chemically modified wood fibers using FTIR spectroscopy for biocomposites [J]. Journal of Applied Polymer Science, 116 (6): 3212-3219.

HAN J, ZHOU C, WU Y, et al. 2013. Self-assembling behavior of cellulose nanoparticles during freeze-drying: Effect of suspension concentration, particle size, crystal structure, and surface charge [J]. Biomacromolecules, 14 (5): 1529-1540.

HAO Y T, XIA Q K, YANG X Z. 2007. Water in minerals from peridotite xenoliths of Hannuoba basalts, Hebei, China: Micro-FTIR results [J]. Acta Mineralogica Et Petrologica, 26 (2): 130-140.

HARTLEY I D, SCHNEIDER M H. 1993. Water vapor diffusion and adsorption characteristics of sugar maple (Acer saccharum, Marsh.) wood polymer composites [J]. Wood Science and Technology, 27 (6): 421-427.

HAXAIRE K, MARECHAL Y, MILAS M, et al. 2003. Hydration of polysaccharide hyaluronan observed by IR spectrometry. I. Preliminary experiments and band assignments [J]. Biopolymers, 72 (1): 10-20.

HENDRICKX R, FERREIRA E S B, BOON J J, et al. 2017. Distribution of moisture in reconstructed oil paintings on canvas during absorption and drying: a neutron radiography and NMR study [J]. Studies in Conservation, 62 (7): 393-409.

HERNANDEZ R E, CACERES C B. 2010. Magnetic resonance microimaging of liquid water distribution in sugar maple wood below fiber saturation point [J]. Wood and Fiber Science, 42 (3): 259-272.

HOFSTETTER K, HINTERSTOISSER B, SALMEN L. 2006. Moisture uptake in native cellulose - the roles of different hydrogen bonds: a dynamic FTIR study using Deuterium exchange [J]. Cellulose, 13 (2): 131-145.

HUNTER A J. 1995. The distinction between average enthalpy and the isosteric heat of water sorbed on wood; and the spatial distribution of the specific enthalpy of the sorbed water [J]. Wood Science and Technology, 29 (5): 377-383.

ISA A, MINAMINO J, MIZUNO H, et al. 2013. Increased water resistance of bamboo Flour/ Polyethylene composites [J]. Journal of Wood Chemistry and Technology, 33 (3): 208-216.

ISHII D, SAITO T, ISOGAI A. 2011. Viscoelastic evaluation of average length of cellulose nanofibers prepared by TEMPO-mediated oxidation [J]. Biomacromolecules, 12 (3): 548-550.

ISLAM M N, KHAN M A, ALAM M K, et al. 2003. Study of water absorption behavior in wood plastic composites by using neutron radiography techniques [J]. Polymer-Plastics Technology and Engineering, 42 (5): 925-934.

ISOGAI T, SAITO T, ISOGAI A. 2011. Wood cellulose nanofibrils prepared by TEMPO electro-mediated oxidation [J]. Cellulose, 18 (2): 421-431.

ITO Y, NAKASHIMA S. 2002. Water distribution in low-grade siliceous metamorphic rocks by micro-FTIR and its relation to grain size: A case from the Kanto Mountain region, Japan [J]. Chemical Geology, 189 (1-2): 1-18.

KACHRIMANIS K, NOISTERNIG M F, GRIESSER U J, et al. 2006. Dynamic moisture sorption and desorption of standard and silicified microcrystalline cellulose [J]. European Journal of Pharmaceutics and Biopharmaceutics, 64 (3): 307-315.

KAMMER M, HEDRICH R, EHRLICH H, et al. 2010. Spatially resolved determination of the structure and composition of diatom cell walls by Raman and FTIR imaging [J]. Analytical and Bioanalytical Chemistry, 398 (1): 509-517.

KHALI D P, RAWAT S P S. 2000. Clustering of water molecules during adsorption of water in brown rot decayed and undecayed wood blocks of Pinus sylvestris [J]. Holz Als Roh- Und Werkstoff, 58 (5): 340-341.

KONDO T. 1997. The assignment of IR absorption bands due to free hydroxyl groups in cellulose [J]. Cellulose, 4 (4): 281-292.

KRUEGER P, MARKOETTER H, HAUSSMANN J, et al. 2011. Synchrotron X-ray tomography for investigations of water distribution in polymer electrolyte membrane fuel cells [J]. Journal of Power Sources, 196 (12): 5250-5255.

LI T Q, HENRIKSSON U, ERIKSSON J C, et al. 1992. Water-cellulose interaction in wood pulp fiber suspensions studied by oxygen-17 and deuterium NMR relaxation. The effect of beating [J]. Langmuir, 8 (2): 680-686.

LIU C, SHAO Z, WANG J, et al. 2016. Eco-friendly polyvinyl alcohol/cellulose nanofiber-Li^+ composite separator for high-performance lithium-ion batteries [J]. RSC Advances, 6 (100): 97912-97920.

LIU Y, YANG Z, DESYATERIK Y, et al. 2008. Hygroscopic behavior of substrate-deposited particles studied by micro-FTIR spectroscopy and complementary methods of particle analysis [J]. Analytical Chemistry, 80 (3): 633-642.

MANHAS N, BALASUBRAMANIAN K, PRAJITH P, et al. 2015. PCL/PVA nanoencapsulated reinforcing fillers of steam exploded/autoclaved cellulose nanofibrils for tissue engineering applications [J]. RSC Advances, 5 (31): 23999-24008.

MANNES D, SONDEREGGER W, HERING S, et al. 2009. Non-destructive determination and quantification of diffusion processes in wood by means of neutron imaging [J]. Holzforschung,

63 (5): 589-596.

MAO Z, YU H, WANG Y, et al. 2014. States of water and pore size distribution of cotton fibers with different moisture ratios [J]. Industrial and Engineering Chemistry Research, 53 (21): 8927-8934.

MENON R S, MACKAY A L, HAILEY J R T, et al. 1987. An NMR determination of the physiological water distribution in wood during drying [J]. Journal of Applied Polymer Science, 33 (4): 1141-1155.

NAKANO T. 1999. Analysis of creep of wood during water adsorption based on the excitation response theory [J]. Journal of Wood Science, 45: 19-23.

NEWMAN R H, DAVIDSON T C. 2004. Molecular conformations at the cellulose-water interface [J]. Cellulose, 11 (1): 23-32.

OH S Y, YOO D I, SHIN Y, et al. 2005. Crystalline structure analysis of cellulose treated with sodium hydroxide and carbon dioxide by means of X-ray diffraction and FTIR spectroscopy [J]. Carbohydrate Research, 340 (15): 2376-2391.

OHMAE Y, SAITO Y, INOUE M, et al. 2009. Water adsorption process of bamboo heated at low temperature [J]. Journal of Wood Science, 55 (1): 13-17.

OLSSON A M, SALMEN L. 2004. The association of water to cellulose and hemicellulose in paper examined by FTIR spectroscopy [J]. Carbohydrate Research, 339 (4): 813-818.

PEARCE R B, FISHER B J, CARPENTER T A, et al. 1997. Water distribution in fungal lesions in the wood of sycamore, Acer pseudoplatanus, determined gravimetrically and using nuclear magnetic resonance imaging [J]. New Phytologist, 135 (4): 675-688.

ROSENKILDE A, GLOVER P. 2002. High resolution measurement of the surface layer moisture content during drying of wood using a novel magnetic resonance imaging technique [J]. Holzforschung, 56 (3): 312-317.

ROSNER S, RIEGLER M, VONTOBEL P, et al. 2012. Within-ring movement of free water in dehydrating Norway spruce sapwood visualized by neutron radiography [J]. Holzforschung, 66 (6): 751-756.

SAKATA I, UEDA T, MURAKAWA H, et al. 2009. Three-dimensional observation of water distribution in PEFC by neutron CT [J]. Nuclear Instruments and Methods in Physics Research Section A-Accelerators Spectrometers Detectors and Associated Equipment, 605 (1-2): 131-133.

SCHUTTLEFIELD J, AL-HOSNEY H, ZACHARIAH A, et al. 2007. Attenuated total reflection Fourier transform infrared spectroscopy to investigate water uptake and phase transitions in atmospherically relevant particles [J]. Applied Spectroscopy, 61 (3): 283-292.

SHI X, ZHENG Y, WANG G, et al. 2014. pH- and electro-response characteristics of bacterial cellulose nanofiber/sodium alginate hybrid hydrogels for dual controlled drug delivery [J]. RSC Advances, 4 (87): 47056-47065.

TANIGUCHI T, HARADA H, NAKATO K. 1978. Determination of water adsorption sites in wood by a hydrogen-deuterium exchange [J]. Nature, 272 (5650): 230-231.

TONEY M F, HOWARD J N, RICHER J, et al. 1995. Distribution of water molecules at Ag(111)/electrolyte interface as studied with surface X-ray scattering [J]. Surface Science, 335: 326-332.

TSHABALALA M A, DENES A R, WILLIAMS R S. 1999. Correlation of water vapor adsorption behavior of wood with surface thermodynamic properties [J]. Journal of Applied Polymer Science, 73 (3): 399-407.

URAKI Y, MATSUMOTO C, HIRAI T, et al. 2010. Mechanical effect of acetic acid lignin adsorption on Honeycomb-Patterned cellulosic films [J]. Journal of Wood Chemistry and Technology, 30 (4): 348-359.

VOGT B D, SOLES C L, LEE H J, et al. 2005. Moisture absorption into ultrathin hydrophilic polymer films on different substrate surfaces [J]. Polymer, 46 (5): 1635-1642.

WEHBE K, FILIK J, FROGLEY M D, et al. 2013. The effect of optical substrates on micro-FTIR analysis of single mammalian cells [J]. Analytical and Bioanalytical Chemistry, 405 (4): 1311-1324.

YUE Y, ZHOU C, FRENCH A D, et al. 2012. Comparative properties of cellulose nano-crystals from native and mercerized cotton fibers [J]. Cellulose, 19 (4): 1173-1187.

ZHANG X, FENG Y, HUANG C, et al. 2017. Temperature-induced formation of cellulose nanofiber film with remarkably high gas separation performance [J]. Cellulose, 24 (12): 5649-5656.

ZHOU M, LU D, DUNSMUIR J, et al. 2000. Irreducible water distribution in sandstone rock: two phase flow simulations in CT-based pore network [J]. Physics and Chemistry of the Earth, Part A: Solid Earth and Geodesy, 25 (2): 169-174.

ZULUAGA R, PUTAUX J L, CRUZ J, et al. 2009. Cellulose microfibrils from banana rachis: effect of alkaline treatments on structural and morphological features [J]. Carbohydrate Polymers, 76 (1): 51-59.

Chapter 11 Water vapor sorption properties of sulfuric acid treated and TEMPO oxidized cellulose nanofiber films

11.1 Introduction

In the stone age, natural materials (such as stones, skins, feathers, trees, etc.) were mainly used; in the bronze age, bronze ware became the main appliance; later, because of its portability and sharpness, iron ware gradually replaced bronze ware; and in modern civilized society, new polymer materials were gradually invented and prepared by chemical and physical methods, and were given special properties by composition and modification to realize the functionalization and intelligentization of materials. Over the past century, scientists have devoted themselves to synthesizing and preparing new polymer materials. Tens of thousands of polymer materials have been synthesized. However, there are only dozens of polymer materials in practical use. Natural polymers not only have unique and excellent properties, but also will not cause environmental pollution. Therefore, the research on new properties of natural polymers has mushroomed in the last ten years. Among them, cellulose is the most abundant on the earth, the most widespread distribution of a natural polymer, its dehydration of glucose from condensation polymerization.

Cellulose is widely found in the cell walls of green plants, such as wood and leaves, which combine with lignin, hemicellulose and pectin to form the cell walls of plants. Cellulose is also abundant in algae and marine biofilms. In addition, some bacteria are able to synthesize greater aspect ratios and stronger bacterial Cellulose. Cellulose is most abundant in hemp plants (such as jute and sisal) and cotton (Berglund et al. 2016; Rosa et al. 2010; Ruan et al. 1996). It is important to note that natural cellulose contains nano-sized crystalline cellulose and amorphous cellulose, which rely on intramolecular and intermolecular hydrogen bonds and van der Waals forces to maintain the aggregated structure of cellulose macromolecules. For example, in woody plants, a linear cellulose chain is a microfiber, with a typical cross-sectional size of

35nm, which makes it possible for us to obtain nano-cellulose from natural cellulose (a nano-cellulose generally refers to a cellulose of a certain dimension at a nanoscale). Now Cellulose is showing great potential in the field of nanomaterials. Cellulose nanofibers have many outstanding properties such as high mechanical strength, low thermal expansion coefficient, high thermal and chemical durability and film-forming capacity (Gamelas et al. 2015; Sun et al. 2015; Xu et al. 2016). In addition, it also has high optical, electrical, magnetic and rheological properties, and has the characteristics of lightness, degradability, biocompatibility and reproducibility of biological materials, so it shows great application prospects in high performance composites.

In the 1990s, especially in the last ten years, the research on nano-cellulose has become a hotspot in the field of cellulose science. The research of nano-cellulose-based composites is a new field, but its development is very rapid. Nanometer cellulose is a kind of promising nanometer material, which has abundant sources and excellent properties. It should be noted that nano-cellulose can provide high transparency and excellent oxygen barrier, which is a very important performance requirement for food and pharmaceutical packaging materials. The low thermal expansion coefficient, high strength, high modulus and high diaphaneity of nano-cellulose make nano-cellulose an ideal reinforcing material for tight coil alignment technology. In addition, nanofibers are increasingly used in bone tissue repair engineering, and have been considered for soft (cartilage, ligaments, blood vessels, etc.) and hard (orthopaedic and dental) tissue grafts because of their superior mechanical, electrical, and thermal properties (Deng et al. 2017; Manhas et al. 2015; Shi et al. 2014). However, hygroscopic behavior is an inherent characteristic of cellulose nanofiber film which strongly affects its surface behavior and leads to reliability problem. Consequently, the water vapor sorption property needs to be understood prior to the commercial utilization of the cellulose nanofiber film.

As the potential application of cellulose nanofiber film is promising, it attracts increasing attention in many scientific fields, and the preparation methods of the cellulose nanofiber film from various cellulosic sources have been extensively studied for several decades (Frenot and Chronakis 2004; Fukuzumi et al. 2009; Jonoobi et al. 2010; Tibolla et al. 2014; Tonoli et al. 2012). At present, the common preparation methods the preparation methods of nanofibers can be divided into two categories: (A) cellulose nanofiber prepared by acid hydrolysis of native celluloses and successive mechanical agitation of the acid-hydrolyzed residues in water, and (B) cellulose nanofiber prepared by 2, 2, 6, 6-tetramethylpiperidine-1-oxy radical (TEMPO)-mediated oxidation of native celluloses followed by mechanical disintegration of the oxidized celluloses in water (Akira et al. 2011; Fujisawa et al. 2011; Jaušovec et al. 2015; Tsuguyuki et al. 2007). The comprehensive study contrasting the similarities and differences between TEMPO-oxidized cellulose nanofiber (TOCNF) and sulfuric acid treated cellulose nanofiber (SACNF) have been reported (Sun et al. 2015). Among them, the TEMPO selective oxidation method is that the wood pulp is oxidized cellulose after catalytic oxidation, and then dispersed cellulose into deionized water to form cellulose nanofiber suspension. This method can be very effective in the natural wood pulp of the primary fiber in the case of no serious damage from the wood pulp in a completely effective separation. The prepared cellulose nanofibers retain the structural characteristics of natural wood pulp, such as high aspect ratio and small cross section. This method not only has high efficiency, but also has low requirements for reaction equipment, more fine structure, more uniform scale, so it is favored. The acid hydrolysis process degrades the amorphous region of cellulose into small molecules, retains the crystalline part of cellulose, and finally forms nano-cellulose in rods. During the reaction, the properties of cellulose nanofibers are affected by the type of acid, hydrolysis environment, time and concentration. For example, the surface of cellulose nanofibers hydrolyzed by hydrochloric acid is unstable and prone to flocculation, while the surface of cellulose nanofibers hydrolyzed by sulfuric acid is more repulsive and can reduce flocculation, so the solution of cellulose nanofibers is more stable. Therefore, sulfuric acid hydrolysis is generally used to prepare cellulose nanofibers instead of hydrochloric acid hydrolysis. Moreover, there are some representative publications available which have provided lots of useful information about these two cellulose nanofibers. It has been proven that the structure of cellulose nanofiber differ from the normal cellulose which can directly influent water vapor sorption behavior (Wang et al. 2006). Due to the complexity of water vapor sorption behavior, there are still many issues that deserve more detailed investigation in the case of TOCNF and SACNF films.

Water vapor sorption is a key property of the cellulose material (Angkuratipakorn et al. 2017; Peresin et al. 2010; Wan et al. 2009; Watanabe et al. 2006). At very low relative humidity in the air, cellulose due to the presence of hydroxyl surface, it can be combined with water molecules, so cellulose has a certain moisture absorption. However, when the relative humidity of air is high, the multilayer adsorption of water molecules will occur due to the existence and action of hydrogen bonds in cellulose pores. When the air humidity rises to a certain degree, a large number of water molecules will be condensed and adsorbed in the nanometer pores, and the adsorption heat will be similar to the liquefaction energy of water. And it has been studied via a number of experimental approaches, such as differential scanning calorimetry (Agrawal et al. 2004; Mccrystal et al. 1997; Mccrystal et al. 1999; Nakamura et al. 1981; Nelson 2010), dynamic vapor sorption (Driemeier et al. 2012; Hill et al. 2010; Xie et al. 2011; Zaihan et al. 2009), dielectric relaxation spectroscopy (Smith 2011; Sugimoto et al. 2008), Fourier transform infrared spectroscopy (Murphy and Pinho 1995; Olsson and Salmén 2004), and nuclear magnetic resonance spectroscopy(Carles and Scallan 2010; Felby et al. 2008; Ogiwara et al. 1969; Topgaard and Söderman 2001). Of these experimental approaches, dynamic vapor sorption (DVS) is an important quantitative method. The isothermal adsorption of nano-cellulose at a certain range of relative humidity has been widely used and a large number of water adsorption data have been provided (Glass et al. 2018; Hill et al. 2012). Using this approach, sorption isotherm and sorption hysteresis of many cellulose materials including natural fibers (Hill et al. 2010; Robertkohler et al. 2003; Xie et al. 2011), regenerated cellulose (Okubayashi et al. 2010), microcrystalline cellulosic fibers (Kachrimanis et al. 2006; Xie et al. 2011) and wood powder (Madamba et al. 1996) have been analyzed. As this method has been confirmed to be able to give highly reproducible sorption data of tested sample over a wide RH range in real time (Guo et al. 2017; Popescu et al. 2014; Xie et al. 2010), it can be used as a reference method to analyze the water sorption behavior. Moreover, water vapor sorption behavior of these materials has also been accurately fitted by parallel exponential kinetics (PEK) model (Hill et al. 2010; Hill et al. 2010; Kachrimanis et al. 2006; Kohler et al. 2010), as shown in the following equation:

$$MC = MC_0 + MC_1(1-\exp(-t/t_1)) + MC_2(1-\exp(-t-/t_2)) \quad (11.1)$$

Where, MC is real-time moisture content, MC_0 is initial moisture content, MC_1 is moisture content related to fast kinetic process, MC_2 is moisture content linked to slow kinetic process, t_1 is characteristic time for fast kinetic process, and t_2 is characteristic time for slow kinetic process. It is obvious that the PEK model has a double exponential form which represents fast and slow kinetic processes (Hill and Xie 2011; Murr and Lackner 2018; Sharratt et al. 2011). The fast kinetic process is considered to be moisture sorption at the sites of surfaces and amorphous regions of the cellulose material, and the slow kinetic process is proposed to be moisture sorption at newly generated sites (Hill et al. 2010). Both processes can be explained by Kelvin-Voigt model, which has the following equation:

$$\varepsilon = (\sigma_0/E)[-\exp(-\lambda t)] \quad (11.2)$$

Where, ε is strain, σ_0 is applied stress, E is modulus, λ is constant equal to E/η, η is viscosity, and *t* is time. In the analysis of the PEK behavior of cellulose material, the strain is regarded as the volume change of the cellulose material caused by water vapor sorption, and the applied stress is considered to be the swelling pressure affected by the water vapor sorption. In addition, the swelling pressure can be calculated using the following formula:

$$\sigma_0 = -(\rho/M)RT \cdot \ln(p_i/p_f) \tag{11.3}$$

Where, ρ and M are density and molecular weight of water, R is gas constant, T is Kelvin temperature, and p_i and p_f are initial and final water vapor pressure. The PEK and Kelvin-Voigt models can be combined to analyze the water vapor sorption kinetics (Keating et al. 2013).

In this work, we used sulfuric acid treatment and TEMPO to oxidize cellulose nano-fiber films, and then used the DVS technique to study water vapor sorption behavior of two nanocellulose films (i.e., SACNF and TOCNF). Firstly, we collected the sorption data of these two cellulose nanofiber films including the running time, real-time sample mass, target relative humidity (RH), actual RH and real-time temperature to examine the differences in the equilibrium moisture content, sorption hysteresis and sorption kinetics. Secondly, the sorption kinetics of these two cellulose nanofiber films was analyzed using PEK model. Finally, Kelvin-Voigt model was introduced to interpret the PEK fitting data and estimate modulus of these two cellulose nanofiber films.

11.2 Material and methods

11.2.1 Materials

This work was focused on two typical cellulose nanofibers: SACNF and TOCNF. For SACNF, dried microfibrillated cellulose (manufactured by Shizuoka University, Japan) was raw material, which was hydrolyzed-pretreated with 48 wt % sulfuric acid for 2h and then was filtered under vacuum. The filtered material was washed with distilled water and then centrifuged at 25 ℃. This process was repeated for several cycles until the pH value of the obtained suspension was neutral. This suspension was then put into a high-pressure homogenizer (Microfluidizer M-110EH-30, Microfluidics Corp., USA) and sheared for 20 cycles. After the homogenization, the concentration of SACNF suspension was adjusted to 0.1 wt % by adding water. Meanwhile, for preparing the TOCNF, the same raw material was oxidized using TEMPO, NaBr and

NaClO solutions for 3h. In this oxidation process, NaClO was added drop wisely, and pH value of solutions was kept within 10.0–10.2 by dripping 1 mol/L sodium hydroxide solution. Then the oxidized material was filtered and washed to neutral. Subsequent preparation procedures were the same as those for the SACNF. The obtained suspensions of these two nanocellulose samples were then dried in culture dish at 40 ℃ for about 5 h to form films. The final film with the diameters of 120 mm and the thickness of about 20 μm was placed between two pieces of cover slips and then sealed in plastic bags. In addition, conductometric titration was used to determine the carboxyl content of TOCNF film according to the literature. This measurement was repeated three times, and the average of carboxyl content was 0.93 mmol/g.

11.2.2 DVS apparatus

A DVS apparatus (DVS AdvantagePlus, Surface Measurement Systems Ltd, London, United Kingdom) was used to analyze dynamic sorption behavior of these two cellulose nanofiber films. In the analyses, the sample masses of SACNF and TOCNF films were 8.9 mg and 8.8 mg, separately. The RH was set to change from 0% to 95% in 5% steps and then decrease to 0%. Real-time sample mass were collected every 30 s. At every step, the RH was kept constant for some time and then increased to the next step as the sample mass decreased at 0.002%/min over a period of 10 min. Data such as the running time, real-time sample mass, and RH were all collected at a constant temperature of 25 ℃ during water sorption process. Remarkably, the sorption isotherms of these two cellulose nanofiber films were recorded as least three times in the preliminary studies. Based on these studies on the reproducibility, the DVS apparatus was confirmed to be able to give reproducible sorption data for these materials and one test was acceptable. The moisture content was calculated using real-time sample mass and the following equation:

$$\text{MC} = \frac{m_2 - m_1}{m_1} \times 100\% \tag{11.4}$$

Where, MC is real-time moisture content, m_1 is dry sample mass, and m_2 is real-time sample mass at a specific RH.

11.2.3 X-ray diffraction (XRD)

Bruker/Siemens Hi-star 2D diffractometer (Bruker AXS, Madison, WI, USA) was used to measure X-ray diffraction (XRD) patterns for TOCNF and SACNF films. The experiments were performed at a scanning rate of 2°/min from 10° to 40° with Cu-Kα radiation generated at 45 kV and 40 mA. The crystallinity index (CI) was calculated by the following equation according to the Segal method.

$$CI = \frac{I_{200} - I_{am}}{I_{200}} \times 100\% \qquad (11.5)$$

where, I_{200} is the maximum intensity of the principal peak, and I_{am} is the intensity of diffraction attributed to amorphous cellulose.

11.2.4 Modulus measurement

Instron 5865 universal material testing apparatus (Instron Engineering Corporation, Norwood, MA, USA) equipped with a 500N load cell load cell and a pair of miniature tensile grips was used to determine the modulus of TOCNF and SACNF films according to the modified ASTM D638-03 standard. Three specimens, 10 mm (width) ×40 mm (length) ×20 μm (thickness) were conditioned at 50% RH and 25℃ for one week. A crosshead speed of 1 mm/min and a gauge length of 8 mm were used for the measurement. The Modulus of TOCNF and SACNF films was obtained from the slope of their stress–strain curves.

11.3 Results and discussion

11.3.1 Water vapor sorption behavior

The running time is one of the most important data collected during the water vapor sorption process. During the sorption experiment, after the target RH was changed into a new value, there existed a delay time of about 4-10 min when the actual RH reached the new setting target RH. After the typical delay time, the actual RH would be maintained while the actual RH fluctuated less than 0.1% in the following time (as shown in Figure 11.1). Meanwhile, the real-time moisture content of the cellulose nanofiber film produced an asymptotic curve against time, and it would approach the equilibrium moisture content (EMC) at this setting target RH. Once the fluctuation of real-time mass was less than 0.002% per min during 10 min, target RH would become next new setting. Here, total running time consisted of the running time spent in one adsorption process from 0% to 95% RH and that needed in the subsequent desorption process from 95% to 0% RH. Meanwhile, it was found that the temperature values were very stable (25 ℃).

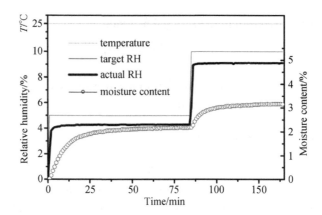

Figure 11.1 Typical changes of the target RH, actual RH and real-time moisture content during the water sorption run

Figure 11.2 represents the measured moisture content of SACNF and TOCNF films during the water sorption process, which includes total running times for these two cellulose nanofiber films. Although the values were very close, the total running times for two cellulose nanofiber films increased in the following order: SACNF<TOCNF. The total running time can be influenced by a number of factors, including maximum water uptake, experimental set up, sample mass, and so on. In the present case, the value of the total running time indicated that SACNF film had lower sorption capacity than TOCNF film. The relatively lower sorption capacity of SACNF could be due to its less amorphous nature.

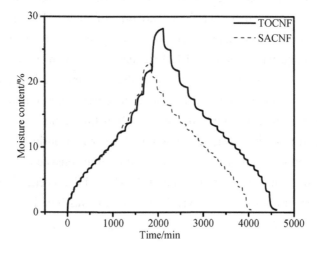

Figure 11.2 Moisture content of TOCNF and SACNF films during the sorption process

Determining equilibrium moisture content (EMC) is a means of evaluating sorption properties of cellulose nanofiber film. Figure 11.3 shows the EMC of SACNF and TOCNF films over a full set range of RH levels. These two films generated sigmoidal isotherm curves during the adsorption and desorption processes. Similar EMC isotherms were observed in other lignocellulosic materials. As the RH was above 80%, there was a rapid increase of the EMC isotherms, which was also found in cellulose whiskers and microfibrils films (Belbekhouche et al. 2011). This phenomenon may be due to the capillary condensation. Moreover, there were obvious differences in EMC values between SACNF and TOCNF films. At the highest RH of 95%, SACNF obtained an EMC of 22.8%, while TOCNF had a higher EMC of 28.2%. This difference may be explained by the fact that SACNF film had higher crystallinity and less amorphous regions, compared with TOCNF film. The obtained X-ray diffraction patterns of SACNF and TOCNF films in this experiment were used to determine crystalline structures, and the CI values of SACNF and TOCNF films were 70.3% and 74.2%, respectively, which were in a good agreement with the results from some published papers. During the preparation of SACNF, the acid hydrolysis treatment (48 wt% H_2SO_4) was carried out for the raw material, which can remove amorphous regions effectively (Isogai et al. 2011). It is clear that the preparation procedure of cellulose film can play a significant role in the water sorption.

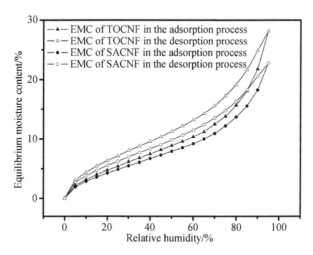

Figure 11.3   Equilibrium moisture content of TOCNF and SACNF films at each given RH level

11.3.2   Sorption hysteresis

The sorption hysteresis curve can be obtained by subtracting EMC in desorption isotherm from that in adsorption isotherm. Figure 11.4 represents the sorption hysteresis curve plotted as function of RH for SACNF and TOCNF films.

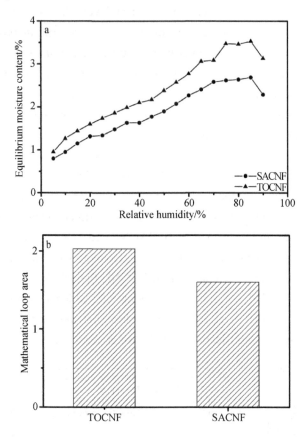

Figure 11.4 Sorption hysteresis (a) and mathematical loop area (b) plotted as a function of RH for TOCNF and SACNF films

To analyze the sorption hysteresis, the mathematical area of the isotherm loops was used to characterize the total hysteresis. It was found that the calculated mathematical area of the isotherm loops for TOCNF film was larger than that for SACNF film. This confirmed that the total sorption hysteresis of TOCNF film were greater than that of SACNF film. As the phenomenon of hysteresis is linked to the ability of the matrix to deform in response to the ingress or egress of water molecules into or out of the material, these differences in the overall sorption hysteresis may be explained by the swelling of cellulosic materials (Bedane et al. 2015). It has been considered that that the ability of water to cause swelling depends on the amorphous content of cellulosic materials. As mentioned before, the amorphous content of TOCNF film is more than that of SACNF film. Hence, the extent of the swelling of TOCNF film should be greater which may account for this variability in the total hysteresis.

To further analyze the sorption hysteresis, the incremental increment and decrement of EMC vs. RH are demonstrated in Figure 11.5. Both SACNF and TOCNF films exhibited greater incremental increment and decrement of EMC at the upper end of RH range than those in other regions. Meanwhile, the incremental increment and decrement of EMC for TOCNF were greater than those for SACNF. All these differences are attributed to the differences in the crystalline structure.

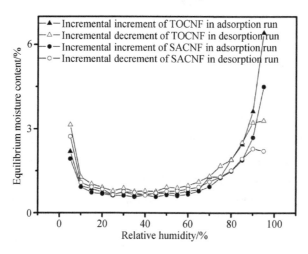

Figure 11.5 Incremental increment and decrement of EMC as a function of RH

11.3.3 Sorption kinetics

Figure 11.6 shows the average rate of sorption at various RHs during adsorption and desorption, in which the average rate at the specific RH is acquired by dividing the increment or decrement of EMC by time. The shape of the measured average rate for both SACNF and TOCNF films was "W". This suggested that the average sorption rates at both low and high RH ranges were higher than those in the middle RH range. This suggested that the average sorption rate was fast at both low and high RH ranges; while, in the middle RH range during adsorption and desorption, the average sorption rate was slow.

In order to further understand the sorption kinetic, the kinetic curves were analyzed using the PEK model. It was important to note that the PEK model could correctly described the sorption kinetics data, as the coefficient of determination (R^2) was higher than 0.99. Two PEK fitting parameters such as characteristic time for fast

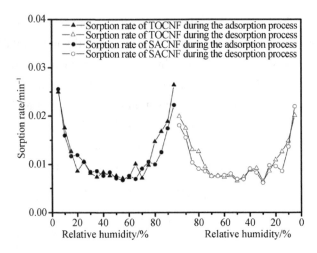

Figure 11.6　Average sorption rates of TOCNF and SACNF films at each given RH

kinetic process (t_1) and characteristic time for slow kinetic process (t_2) are given in Figure 11.7(a) and 11.7(b). It was shown that t_1 and t_2 varied during the adsorption and desorption processes, especially at high and low RH; this variation in t_1 and t_2 was also observed in microcrystalline cellulose. Moreover, there was an obvious numerical difference between the adsorption and desorption characteristic times for the slow kinetic processes (t_2), while the numerical difference between the adsorption and desorption characteristic times for the fast kinetic processes was not so clear (t_1). All these suggested that the adsorption and desorption processes were asymmetric processes. The PEK fitting parameters MC_1 and MC_2 were also obtained, and cumulative moisture contents linked to two kinetic processes (cumulative MC_1 and MC_2) as a function of RH are displayed in Figure 11.7(c) and 11.7(d). It was found that both SACNF and TOCNF films had more cumulative MC_1 than cumulative MC_2, which confirmed that the fast kinetic process was the dominant process for controlling the adsorbed moisture content throughout the hygroscopic range. The same conclusion has been reported in the measurements of cotton linter and microcrystalline cellulose. In addition, for SACNF and TOCNF films, at the highest RH of 95%, both cumulative MC_1 and cumulative MC_2 increased in the order of SACNF < TOCNF [(Figure 11.7(c) and 11.7(d)]. As mentioned before, the fast kinetic process is related to the readily accessible sorption sites of the surfaces and amorphous regions, while the slow kinetic process is linked to the newly generated sorption sites. Therefore, the increase of the cumulative MC_1 may be due to the increase of hydroxyl groups of the surfaces and amorphous regions whose growth order is SACNF < TOCNF confirmed. Meanwhile, the cumulative MC_2 in this study may be mainly attributed to the swelling which can generate new sorption sites. The extents of swelling of these two nanocellulose films were proved to increase in the same order.

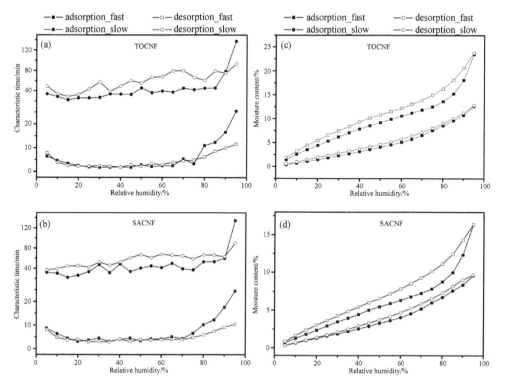

Figure 11.7 Characteristic times t_1 and t_2 of TOCNF (a) and SACNF (b) films and cumulative moisture contents MC_1 and MC_2 of TOCNF (c) and SACNF (d) films at different RH levels.

11.3.4 The applicability of the Kelvin-Voigt model

On this basis of the PEK data, using the Kelvin-Voigt model, the modulus of these two cellulose nanofiber films was calculated. This calculation method has been previously applied in normal cellulose material, and here it was introduced for these two cellulose nanofiber films. As shown previously, the kinetics of the adsorption and desorption processes consist of a fast and a slow kinetic process, which has its own modulus (E_1 and E_2) (Kohler et al. 2010; Popescu et al. 2015). E_1 is the modulus associated with the fast kinetic process, and E_2 is the modulus linked to the slow kinetic process. E_1 and E_2 of SACNF and TOCNF films are shown in Figure 11.8.

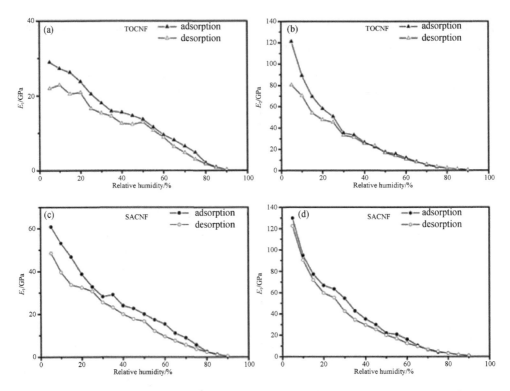

Figure 11.8 E_1 of TOCNF (a) and SACNF (b) films and E_2 of TOCNF (c) and SACNF (d) films at different RH levels during sorption process

As shown in this figure, the values of E_1 and E_2 decreased with increasing RH, which was may be due to the effect of absorbed water on the plasticization of cellulose nanofiber film. What's more, at the RH of 50%, the modulus values of (19.7±3.0) GPa and (15.2±2.3) GPa were calculated for SACNF and TOCNF films. As no study has been reported using Kelvin-Vogit model to calculate the modulus of SACNF and TOCNF films so far, the comparison of the modulus obtained by Kelvin-Vogit model and other methods is necessary. Here, Instron 5865 universal material testing apparatus was used to determine the modulus as described before, and the modulus values of SACNF and TOCNF films were (8.8±1.9) GPa and (8.2±1.8) GPa at the RH of 50%, which were very close to the reported values (Barbash et al. 2016; Fujisawa et al. 2011). It was clearly that the modulus value calculated using Kelvin-Vogit model was higher than that determined by tension method. Meanwhile, the modulus determined using tension method were obtained by applying an external pressure on the cellulose nanofiber film, while those calculated using Kelvin-Vogit model were based on the application of the internal pressure. Although the testing mechanism was different, the obtained modulus values were of the same order of magnitude. Meanwhile, some other research reported that the modulus value of SACNF was between 2.1 GPa and 36.0 GPa and the modulus value of TOCNF was between 4.4 GPa and 25.8 GPa under the ordinary condition of 50% RH (Eichhorn 2012; Rodionova et al. 2012; Shinichiro et al. 2011). It was noteworthy that the modulus calculated using the Kelvin-Vogit model was in the range of modulus values quoted in the literature.

11.4 Conclusions

By using DVS apparatus, a detailed investigation on water vapor sorption behavior of two cellulose nanofiber films (i.e., SACNF and TOCNF) was achieved during the sorption process. Some important data, such as the running time, real-time sample mass, target RH, actual RH and real-time temperature were obtained, and it provided convenient condition for comparative analysis of the total running time, equilibrium moisture content, sorption hysteresis and sorption kinetics between the SACNF and TOCNF film. It was important to note that SACNF and TOCNF had EMC values of 22.8% and 28.2% at 95% RH, respectively. Meanwhile, sorption kinetics data were fitted using PEK model. The PEK fitting parameters suggested that the numerical difference between the adsorption and desorption characteristic times for the slow kinetic processes was obvious and cumulative moisture contents related to two kinetic processes increased in the order of SACNF < TOCNF. Furthermore, two Kelvin-Voigt elements were introduced to interpret the PEK behavior and calculate the modulus of SACNF and TOCNF films. It is important to note that SACNF and TOCNF films have modulus values of (19.7±3.0) GPa and (15.2±2.3) GPa, and these calculated using Kelvin-Vogit model in this study are in the range of modulus values quoted in the literature.

References

AGRAWAL A M, MANEK R V, KOLLING W M, et al. 2004. Water distribution studies within microcrystalline cellulose and chitosan using differential scanning calorimetry and dynamic vapor sorption analysis [J]. J Pharm Sci, 93 (7): 1766-1779.

AKIRA I, TSUGUYUKI S, HAYAKA F. 2011. TEMPO-oxidized cellulose nanofibers [J]. Nanoscale, 3 (1): 71-85.

ANGKURATIPAKORN T, SINGKHONRAT J, CHRISTY A A. 2017. Comparison of water adsorption properties of cellulose and cellulose nanocrystals studied by Near-Infrared spectroscopy and gravimetry [J]. Key Engineering Materials, 735: 235-239.

BARBASH V A, YASCHENKO O V, ALUSHKIN S V, et al. 2016. The effect of mechanochemical treatment of the cellulose on characteristics of nanocellulose films [J]. Nanoscale Research Letters, 11 (1): 410-417.

BEDANE A H, EIĆ M, FARMAHINI-FARAHANI M, et al. 2015. Water vapor transport properties of regenerated cellulose and nanofibrillated cellulose films [J]. Journal of Membrane Science, 493: 46-57.

BELBEKHOUCHE S, BRAS J, SIQUEIRA G, et al. 2011. Water sorption behavior and gas barrier properties of cellulose whiskers and microfibrils films [J]. Carbohydrate Polymers, 83 (4): 1740-1748.

BERGLUND L, NOËL M, AITOMÄKI Y, et al. 2016. Production potential of cellulose nanofibers from industrial residues: Efficiency and nanofiber characteristics [J]. Industrial Crops and Products, 92: 84-92.

CARLES J E, SCALLAN A M. 2010. The determination of the amount of bound water within cellulosic gels by NMR spectroscopy [J]. Journal of Applied Polymer Science, 17 (6): 1855-1865.

DENG Z, JUNG J, ZHAO Y. 2017. Development, characterization, and validation of chitosan adsorbed cellulose nanofiber (CNF) films as water resistant and antibacterial food contact packaging [J]. LWT - Food Science and Technology, 83: 132-140.

DRIEMEIER C, MENDES F M, OLIVEIRA M M. 2012. Dynamic vapor sorption and thermoporometry to probe water in celluloses [J]. Cellulose, 19 (4): 1051-1063.

EICHHORN S J. 2012. Stiff as a board: Perspectives on the crystalline modulus of cellulose [J]. ACS Macro Letters, 1 (11): 1237-1239.

FELBY C, THYGESEN L G, KRISTENSEN J B, et al. 2008. Cellulose-water interactions during enzymatic hydrolysis as studied by time domain NMR [J]. Cellulose, 15 (5): 703-710.

FRENOT A, CHRONAKIS I S. 2004. Polymer nanofibers assembled by electrospinning [J]. Current Opinion in Colloid and Interface Science, 8 (1): 64-75.

FUJISAWA S, OKITA Y, FUKUZUMI H. 2011. Preparation and characterization of TEMPO-oxidized cellulose nanofibril films with free carboxyl groups [J]. Carbohydrate Polymers, 84 (1): 579-583.

FUKUZUMI H, SAITO T, IWATA T. 2009. Transparent and high gas barrier films of cellulose nanofibers prepared by TEMPO-Mediated oxidation [J]. Biomacromolecules, 10 (1): 162-165.

GAMELAS J A F, PEDROSA J, LOURENÇO A F, et al. 2015. On the morphology of cellulose nanofibrils obtained by TEMPO-mediated oxidation and mechanical treatment [J]. Micron, 72: 28-33.

GLASS S V, BOARDMAN C R, THYBRING E E, et al. 2018. Quantifying and reducing errors in equilibrium moisture content measurements with dynamic vapor sorption (DVS) experiments [J]. Wood Science and Technology, 52 (4): 909-927.

GUO X, WU Y, XIE X. 2017. Water vapor sorption properties of cellulose nanocrystals and nanofibers using dynamic vapor sorption apparatus [J]. Scientific Reports, 7 (1): 14207.

HILL C A S, NORTON A J, NEWMAN G. 2010. The water vapour sorption properties of Sitka spruce determined using a dynamic vapour sorption apparatus [J]. Wood Science and Technology, 44 (3): 497-514.

HILL C A S, NORTON A J, NEWMAN G. 2010. The water vapour sorption properties of Sitka spruce determined using a dynamic vapour sorption apparatus [J]. Wood Science and Technology, 44 (3): 497-514.

HILL C A S, NORTON A, NEWMAN G. 2010. Analysis of the water vapour sorption behaviour of Sitka spruce [Picea sitchensis (Bongard) Carr.] based on the parallel exponential kinetics model [J]. Holzforschung, 64 (4): 469-473.

HILL C A S, NORTON A, NEWMAN G. 2010. The water vapor sorption behavior of flax fibers-analysis using the parallel exponential kinetics model and determination of the activation energies of sorption [J]. Journal of Applied Polymer Science, 116 (4): 2166-2173.

HILL C A S, RAMSAY J, KEATING B, et al. 2012. The water vapour sorption properties of thermally modified and densified wood [J]. Journal of Materials Science, 47 (7): 3191-3197.

HILL C A S, XIE Y. 2011. The dynamic water vapour sorption properties of natural fibres and viscoelastic behaviour of the cell wall: Is there a link between sorption kinetics and hysteresis? [J]. Journal of Materials Science, 46 (11): 3738-3748.

ISOGAI A, SAITO T, FUKUZUMI H. 2011. TEMPO-oxidized cellulose nanofibers [J]. Nanoscale, 3 (1): 71-85.

JAUŠOVEC D, VOGRINČIČ R, KOKOL V. 2015. Introduction of aldehyde vs. Carboxylic groups to cellulose nanofibers using laccase/TEMPO mediated oxidation [J]. Carbohydrate Polymers, 116: 74-85.

JONOOBI M, HARUN J, MATHEW A P, et al. 2010. Mechanical properties of cellulose nanofiber (CNF) reinforced polylactic acid (PLA) prepared by twin screw extrusion [J]. Composites Science and Technology, 70 (12): 1742-1747.

KACHRIMANIS K, NOISTERNIG M F, GRIESSER U J, et al. 2006. Dynamic moisture sorption and desorption of standard and silicified microcrystalline cellulose [J]. European Journal of Pharmaceutics and Biopharmaceutics, 64 (3): 307-315.

KEATING B A, HILL C A S, SUN D, et al. 2013. The water vapor sorption behavior of a galactomannan cellulose nanocomposite film analyzed using parallel exponential kinetics and the Kelvin – Voigt viscoelastic model [J]. Journal of Applied Polymer Science, 129 (4): 2352-2359.

KOHLER R, ALEX R, BRIELMANN R, et al. 2010. A new kinetik model for water sorption isotherms of cellulosic materials [J]. Macromolecular Symposia, 244 (1): 89-96.

MADAMBA P S, DRISCOLL R H, BUCKLE K A. 1996. The thin-layer drying characteristics of garlic slices [J]. Journal of Food Engineering, 29 (1): 75-97.

MANHAS N, BALASUBRAMANIAN K, PRAJITH P, et al. 2015. PCL/PVA nanoencapsulated reinforcing fillers of steam exploded/autoclaved cellulose nanofibrils for tissue engineering applications [J]. RSC Advances, 5 (31): 23999-24008.

MCCRYSTAL C B, FORD J L, RAJABISIAHBOOMI A R. 1997. A study on the interaction of water and cellulose ethers using differential scanning calorimetry [J]. Thermochimica Acta, 294 (294): 91-98.

MCCRYSTAL C B, FORD J L, RAJABI-SIAHBOOMI A R. 1999. Water distribution studies within cellulose ethers using differential scanning calorimetry. 1. Effect of polymer molecular weight and drug addition [J]. Journal of Pharmaceutical Sciences, 88 (8): 792-796.

MURPHY D, PINHO M N D. 1995. An ATR-FTIR study of water in cellulose acetate membranes prepared by phase inversion [J]. Journal of Membrane Science, 106 (3): 245-257.

MURR A, LACKNER R. 2018. Analysis on the influence of grain size and grain layer thickness on the sorption kinetics of grained wood at low relative humidity with the use of water vapour sorption experiments [J]. Wood Science and Technology, 52 (3): 753-776.

NAKAMURA K, HATAKEYAMA T, HATAKEYAMA H. 1981. Studies on bound water of cellulose by differential scanning calorimetry [J]. Textile Research Journal, 51 (9): 607-613.

NELSON R A. 2010. The determination of moisture transition in cellulosic materials using differential scanning calorimetry [J]. Journal of Applied Polymer Science, 21 (3): 645-654.

OGIWARA Y, KUBOTA H, HAYASHI S, et al. 1969. Studies of water adsorbed on cellulosic materials by a high resolution NMR spectrometer [J]. Journal of Applied Polymer Science, 13 (8): 1689-1695.

OKUBAYASHI S, GRIESSER U J, BECHTOLD T. 2010. Moisture sorption/desorption behavior of various manmade cellulosic fibers [J]. Journal of Applied Polymer Science, 97 (4): 1621-1625.

OLSSON A M, SALMÉN L. 2004. The association of water to cellulose and hemicellulose in paper examined by FTIR spectroscopy. [J]. Carbohydrate Research, 339 (4): 813-818.

PERESIN M S, YOUSSEF H, ARJA-HELENA V, et al. 2010. Effect of moisture on electrospun nanofiber composites of poly(vinyl alcohol) and cellulose nanocrystals [J]. Biomacromolecules, 11 (9): 2471-2477.

POPESCU C M, HILL C A S, ANTHONY R, et al. 2015. Equilibrium and dynamic vapour water sorption properties of biochar derived from apple wood [J]. Polymer Degradation and Stability, 111: 263-268.

POPESCU C M, HILL C A S, CURLING S, et al. 2014. The water vapour sorption behaviour of acetylated birch wood: How acetylation affects the sorption isotherm and accessible hydroxyl content [J]. Journal of Materials Science, 49 (5): 2362-2371.

ROBERTKOHLER, CK R, BERNHARDAUSPERGER, et al. 2003. A numeric model for the kinetics of water vapor sorption on cellulosic reinforcement fibers [J]. Composite Interfaces, 10 (2-3): 255-276.

RODIONOVA G, LENES M, ERIKSEN Ø, et al. 2012. Mechanical and oxygen barrier properties of films prepared from fibrillated dispersions of TEMPO-oxidized Norway spruce and Eucalyptus pulps [J]. Cellulose, 19 (3): 705-711.

ROSA I M D, KENNY J M, PUGLIA D, et al. 2010. Morphological, thermal and mechanical characterization of okra (Abelmoschus esculentus) fibres as potential reinforcement in polymer composites [J]. Composites Science and Technology, 70 (1): 116-122.

RUAN R, LUN Y, ZHANG J, et al. 1996. Structure-function relationships of highly refined cellulose made from agricultural fibrous residues [J]. Applied Engineering in Agriculture, 12 (4): 465-468.

SHARRATT V, HILL C A S, ZAIHAN J, et al. 2011. The influence of photodegradation and weathering on the water vapour sorption kinetic behaviour of scots pine earlywood and latewood [J]. Polymer Degradation and Stability, 96 (7): 1210-1218.

SHI X, ZHENG Y, WANG G, et al. 2014. pH- and electro-responsive characteristics of bacterial cellulose nanofiber sodium alginate hybrid hydrogels for the dual controlled drug delivery [J]. RSC Advances, 4 (87): 47056-47065.

SHINICHIRO I, AKIRA I, TADAHISA I. 2011. Structure and mechanical properties of wet-spun fibers made from natural cellulose nanofibers [J]. Biomacromolecules, 12 (3): 831-836.

SMITH G. 2011. Dielectric analysis of water in microcrystalline cellulose [J]. Pharmacy and Pharmacology Communications, 1 (9): 419-422.

SUGIMOTO H, MIKI T, KANAYAMA K, et al. 2008. Dielectric relaxation of water adsorbed on cellulose [J]. Journal of Non-Crystalline Solids, 354 (27): 3220-3224.

SUN X, WU Q, REN S, et al. 2015. Comparison of highly transparent all-cellulose nanopaper prepared using sulfuric acid and TEMPO-mediated oxidation methods [J]. Cellulose, 22 (2): 1123-1133.

TIBOLLA H, PELISSARI F M, MENEGALLI F C. 2014. Cellulose nanofibers produced from banana peel by chemical and enzymatic treatment [J]. Lwt-Food Science and Technology, 59 (2): 1311-1318.

TONOLI G H D, TEIXEIRA E M, CORRÊA A C, et al. 2012. Cellulose micro/nanofibres from Eucalyptus kraft pulp: Preparation and properties [J]. Carbohydrate Polymers, 89 (1): 80-88.

TOPGAARD D, SÖDERMAN O. 2001. Diffusion of water absorbed in cellulose fibers studied with 1H-NMR [J]. Langmuir, 17 (9): 2694-2702.

TSUGUYUKI S, SATOSHI K, YOSHIHARU N, et al. 2007. Cellulose nanofibers prepared by TEMPO-mediated oxidation of native cellulose [J]. Biomacromolecules, 8 (8): 2485.

WAN, YZ, LUO, et al. 2009. Mechanical, moisture absorption, and biodegradation behaviours of bacterial cellulose fibre-reinforced starch biocomposites [J]. Composites Science and Technology,

69 (7): 1212-1217.

WANG J, KALINICHEV A G, KIRKPATRICK R J. 2006. Effects of substrate structure and composition on the structure, dynamics, and energetics of water at mineral surfaces: A molecular dynamics modeling study [J]. Geochimica Et Cosmochimica Acta, 70 (3): 562-582.

WATANABE A, MORITA S, OZAKI Y. 2006. A study on water adsorption onto microcrystalline cellulose by near-infrared spectroscopy with two-dimensional correlation spectroscopy and principal component analysis [J]. Applied Spectroscopy, 60 (9): 1054-1061.

XIE Y, HILL C A S, JALALUDIN Z, et al. 2011. The dynamic water vapour sorption behaviour of natural fibres and kinetic analysis using the parallel exponential kinetics model [J]. Journal of Materials Science, 46 (2): 479-489.

XIE Y, HILL C A S, JALALUDIN Z, et al. 2011. The water vapour sorption behaviour of three celluloses: Analysis using parallel exponential kinetics and interpretation using the Kelvin-Voigt viscoelastic model [J]. Cellulose, 18 (3): 517-530.

XIE Y, HILL C A S, XIAO Z, et al. 2011. Dynamic water vapour sorption properties of wood treated with glutaraldehyde [J]. Wood science and technology, 45 (1): 49-61.

XIE Y, HILL C A S, XIAO Z, et al. 2010. Water vapor sorption kinetics of wood modified with glutaraldehyde [J]. Journal of Applied Polymer Science, 117 (3): 1674-1682.

XU Z, LI J, ZHOU H, et al. 2016. Morphological and swelling behavior of cellulose nanofiber (CNF)/poly(vinyl alcohol) (PVA) hydrogels: Poly(ethylene glycol) (PEG) as porogen [J]. RSC Advances, 6 (49): 43626-43633.

ZAIHAN J, HILL C, CURLING S, et al. 2009. Moisture adsorption isotherms of acacia mangium and endospermum malaccense using dynamic vapour sorption [J]. Journal of Tropical Forest Science, 21 (3): 277-285.

Chapter 12 Water vapor sorption properties of cellulose nanocrystals and nanofibers using dynamic vapor sorption apparatus

12.1 Introduction

Biomass is the most widespread material on the earth. It mainly includes all plants, microorganisms, animals which take plants and microorganisms as food, and the wastes produced by them, such as plants, crops, forest products, marine products, and even urban wastes such as waste paper, natural fibers, etc. It has the characteristics of rich resources, renewable, low pollution and wide distribution. It is estimated that cellulose and lignin, as the main components of plant biomass, are regenerated at a certain rate every year. If this part of resources is well utilized, human beings have an inexhaustible treasure house of resources. With the depletion of non- renewable resources such as coal, oil and natural gas, and the increasing environmental pollution, the full development and utilization of renewable resources has become an inevitable trend.

In the 1920s and 1940s, cellulose was the core and leading subject of polymer research, followed by research on the industrial application of cellulose and its derivatives. The oil crisis in the 1970s also prompted the focus of research on cellulosic biomass resources. Cellulose is mainly synthesized by plants through photosynthesis. It is the most abundant natural renewable organic macromolecule compound on the earth, and has become an indispensable resource for human society. Cellulose can be extracted from agricultural waste, such as cotton linter, sugarcane bagasse, banana rachis, mulberry bark, soybean pods, wheat, straw, and soy hulls, etc., with low cost, industrial production and high added value (Mandal and Chakrabarty 2011; Ramalho et al. 2013; Suopajärvi et al. 2015; Velásquez-Cock et al. 2016).

Nanotechnology is a brand-new science and technology which was gradually developed in the late 1980s and early 1990s. The development of nanotechnology and the emergence of nanomaterials mark a new level of human understanding of nature, and will have an important impact on industrial upgrading and economic growth in the future. Therefore, nanotechnology has become a priority area of support and development for many developed countries and newly industrialized countries in recent years. Nano-cellulose with excellent properties, such as large specific surface area, high strength, high crystallinity and ultra-fine structure, can be prepared by self-assembly and multifunctional manipulation of cellulose molecules in nano-scale (Gamelas et al. 2015; Malladi et al. 2018; Nechyporchuk et al. 2016; Sun et al. 2016). These special properties can significantly improve the electrical, optical, magnetic, insulating and even superconductivity of existing composites, and reduce the consumption of materials and energy to achieve resource sustainability.

Nano cellulose has been widely used in papermaking, tissue engineering, food industry and green packaging materials (Azeredo et al. 2016; Deng et al. 2017; Ferrer et al. 2017; Li et al. 2015; Manhas et al. 2015; Shi et al. 2014). And the potential uses of nano-cellulose are also becoming increasingly popular in many scientific fields, including materials science, electronics and biomedicine (Aytac et al. 2015; Balea et al. 2017; Dumanli 2017; Hoeng et al. 2016; Koga et al. 2014; Lee et al. 2014; Lu et al. 2014). The hydroxyl group (—OH) on the surface of nano-cellulose is abundant, which makes it have excellent hydrophilicity. The hydrophilicity of nano-cellulose can be directly applied to the preparation of hydrophilic polymer composites, such as polyhydroxyethylene methyl acrylate, polyvinyl alcohol (PVA), epoxy resin, etc. However, these hydrophilic —OH groups make the compatibility between nano-cellulose and hydrophobic polymeric materials poor, resulting in poor interface interaction between nano-cellulose filler and polymer matrix, and poor composite properties. In addition, a large number of hydroxyl (—OH) groups lead to the formation of a large number of hydrogen bonds in and between the molecules of nano-cellulose. The existence of hydrogen bond makes the nano-cellulose easily agglomerated, and the agglomeration phenomenon makes the nano-cellulose unevenly dispersed in the polymer matrix, and the interface interaction between the nano-cellulose and the polymer molecular chain weakens, thus the mechanical properties of the composite materials are reduced. At the same time, hydrophilic nanocellulose adsorbs water under hygrothermal conditions (Newman and Davidson 2004; Uraki et al. 2010), which strongly affects its surface behavior and leads to reliability problems (Isa et al. 2013; Vogt et al. 2005). Consequently, the water sorption behavior of nanocellulose needs to be understood prior to the commercial utilization of nanocellulose.

Nanocellulose refers to rod-like cellulose nanocrystals and needlelike cellulose nanofibers whose structures differ from the normal cellulose (Endes et al. 2016; Isogai 2013; Xu et al. 2013). It has been proven that the structure and composition are important factor for controlling the water sorption behavior. Therefore, the water sorption behavior of nanocellulose is complex and different to that of the normal cellulose materials. However, until recently the water sorption behavior of nanostructured cellulose had not been studied.

Water sorption is a property of cellulose that has been studied by a number of experimental methods (Driemeier et al. 2012; Mccrystal et al. 1997; Olsson and Salmén 2004; Sugimoto et al. 2008; Topgaard and Söderman 2001). Among these methods, the dynamic vapor sorption (DVS) method has obvious advantages over the traditional methods, such as short time consuming, large range of relative humidity, accurate test results and repeatability of data (Guo et al. 2017; Popescu et al. 2014; Xie et al. 2010). Using this method, the sorption isotherms of cellulose materials as well as the extent of sorption hysteresis have been obtained (Agrawal et al. 2004; Driemeier et al. 2012; Kachrimanis et al. 2006; Madamba et al. 1996; Okubayashi et al. 2010; Robertkohler et al. 2003; Xie et al. 2011). The water vapor sorption behavior of many cellulose materials, including natural fibers, regenerated cellulose, microcrystalline cellulosic fibers, and wood powder, have been analyzed by using the parallel exponential kinetics (PEK) model as follows (Hill et al. 2010; Hill et al. 2010; Kachrimanis et al. 2006; Kohler et al. 2010; Okubayashi et al. 2010; Xie et al. 2011).

$$MC = MC_0 + MC_1(1 - \exp(-t/t_1)) + MC_2(1 - \exp(-t/t_2)) \quad (12.1)$$

Where, MC is moisture content, MC_0 is moisture content at the initial time, MC_1 is moisture content related to the fast step, MC_2 is moisture content related to slow step, t_1 is the time to reach equilibrium in the fast step, and t_2 is the time to reach equilibrium in the slow step. Clearly, in the PEK model, the water sorption behavior consists of a fast and slow step (Argyropoulos et al. 2011; Bessadok et al. 2009; Hill and Xie 2011; Murr and Lackner 2018; Sharratt et al. 2011). The fast step is assigned to the inflow of water onto the surface. The slow step is related to new absorption points. Both the steps can be described by the Kelvin-Vogit model as follows:

$$\varepsilon = (\sigma_0/E)[1 - \exp(-t/\varphi)] \quad (12.2)$$

Where, ε is strain, σ_0 is instantaneous stress increase, E is the elastic modulus, φ is a constant equal to η/E, and η is viscosity. The PEK and Kelvin-Voigt models can be combined to analyze the water vapor sorption kinetics (Hill et al. 2012; Hill et al. 2013; Keating et al. 2013; Simon et al. 2017; Xie et al. 2011; Xin et al. 2018).

In this work, we used the dynamic vapor sorption apparatus to study the water vapor sorption of four typical nanocelluloses (i.e., CNC I, CNC II, CNF I, and CNF II). The data concerned running time, real-time mass of samples, the predetermined and actual relative humidity, and sample mass change rate. The obtained data were fitted by PEK model. Then, Kelvin-Voigt model was applied to interpret PEK behavior and calculate the modulus and viscosity of these nanocelluloses.

12.2 Material and methods

12.2.1 Materials

This work was focused on four typical nanocelluloses: CNC I, CNC II, CNF I, and CNF II. The preparation procedure was clearly reported; therefore only a brief description is provided. For CNC I, dried, bleached wood pulp was hydrolyzed for 1 h with 64 wt % sulfuric acid and then was filtered under vacuum. The filtered material was mixed with distilled water thoroughly for 20 min and then centrifuged at 26 °C for three cycles. The suspension was obtained after each washing by centrifugation and dissolution in distilled water. The process was repeated for several days until a pH neutral solution was obtained. This material was then put into a high-pressure homogenizer (Microfluidizer M-110P, Microfluidics Corp., Newton, MA, USA). The concentration of the CNC I suspension was 1.0 wt % and was adjusted to 0.1 wt % by adding water. For preparing CNF I, the concentration of sulfuric acid was changed to 48 wt %, while the other conditions were kept unchanged. To prepare cellulose nanocrystals and nanofibers with cellulose II structures, the raw material was turned into mercerized wood pulp. The obtained suspensions of four nanocellulose samples were then quickly frozen at −75 °C for about 2 h and freeze-dried to form film. The final film was placed between two cover slips and then was sealed in plastic bags.

12.2.2 DVS setup

A DVS apparatus (DVS AdvantagePlus, Surface Measurement Systems Ltd, London, United Kingdom) was used to determine the isotherms and dynamic sorption behavior of the four nanocellulose specimens. The data such as the running time, real-time mass of the samples, and the actual RH at a constant temperature of 25 °C were obtained during the DVS process. The RH was set to change from 0% to 95% in 5% steps and then decrease to 0%. At every stage, the RH was kept constant for some time and then increased to the next stage as the sample mass decreased at 0.002% per minute.

12.3 Results and discussion

12.3.1 Water vapor sorption behavior

The running time is an important parameter for characterizing the water vapor sorption behavior of nanocellulose. As shown in Figure 12.1, after changing the target RH to a new setting, there was typically a delay of approximately 4-10 min before the actual RH reached a steady state. Although the stabilized actual RH was not exactly equal to the target RH, the difference was within ±0.1%. Meanwhile, after changing the target RH, the real-time moisture content of the sample generated an asymptotic curve against time, and it could reach the equilibrium moisture content (EMC) if the stability time was long enough. Once the change of moisture content was less than 0.002%/min after 10 min, the target RH was changed to the next setting.

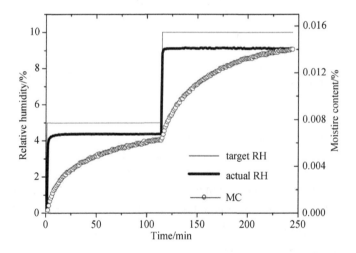

Figure 12.1 Water vapor adsorption behavior of CNC I at the RH of 5% and 10% indicating target RH (narrow line), actual RH (bold line), and moisture content (circle points)

Over the time profile in the isotherm run, the total running times for the four nanocellulose samples are shown in Figure 12.2. Although the values were very close, the total running times for four nanocellulose samples increased. The total running time is related to the water adsorption. So the running time increases with increase in the water adsorption. As shown in Figure 12.2, the total running time of CNC was less than that of CNF in the entire RH range. This is likely due to the lower sorption ability of CNC compared to that of CNF. The relatively high sorption ability of CNF could be due to its high amorphous nature. The total running times of the nanocellulose with cellulose I structure was less than that of the nanocellulose with cellulose II structure throughout the entire RH range. This is probably because increasing the RH over full range increases both the water's accessible area and the swelling of nanocellulose with the cellulose II structure. In addition, as shown in Figure 12.2, the total running time for the desorption process was slightly longer than that for the adsorption process, causing the hysteresis.

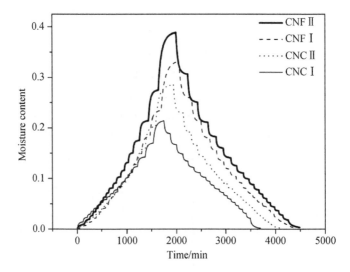

Figure 12.2   The moisture content of four nanocellulose samples at various RHs in the adsorption and desorption processes

EMC is another important parameter for characterizing water vapor sorption. As expected, the EMC of these four cellulose samples, as a function of RH, generated typical sigmoidal isotherms in both the adsorption and desorption steps, as shown in Figure 12.3. Similar EMC isotherms were observed in other normal cellulose materials. As shown in Figure 12.3, the EMC increased rapidly when the RH was above 80%, and this increase was due to the increase in the condensed capillary water. These four cellulose samples at 95% RH had values of 0.2142, 0.2856, 0.3321, and 0.3890, respectively. These values were higher than those of the other cellulose fibers, such as cotton linter and α-cellulose. This was attributed to the larger pore volume and pore diameter of the CNC and CNF films that were made of nanoparticles. The EMC of CNF was higher than that of the CNC owing to the lower degree of crystallinity and the amorphous nature of CNF. In addition, the EMCs of cellulose nanocrystals and nanofibers with cellulose II structures were higher than those with cellulose I structure.

These differences derive from the crystalline structure of cellulose. It is generally known that mercerization modifies the crystalline structure from Ⅰ to Ⅱ. During mercerization, cellulose with cellulose I structure swells, and the cellulose II structure absorbs more water. Han *et al.* showed that the Fourier transform infrared (FTIR)

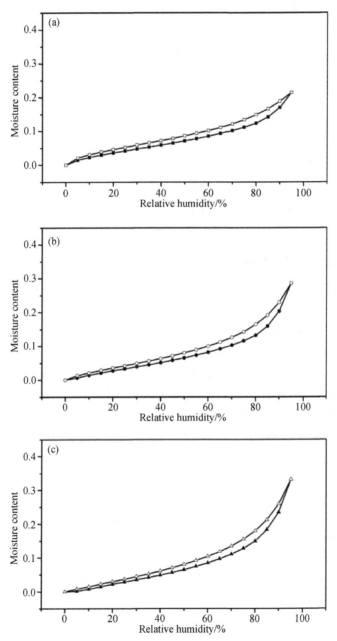

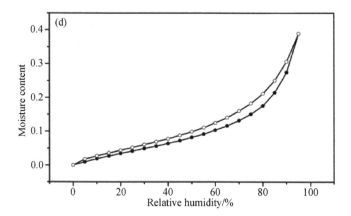

Figure 12.3 Equilibrium moisture content of CNC I (a), CNC II (b), CNF I (c), and CNF II (d) in the adsorption (solid points) and desorption (empty points) steps

spectroscopy signature of OH in cellulose II was higher. All these above suggest that cellulose materials with cellulose II have higher water adsorption than those with cellulose I.

12.3.2 Sorption hysteresis

Figure 12.4 shows the sorption hysteresis over the full RH range in the four nanocellulose samples during adsorption and desorption processes. High sorption hysteresis was clearly observed in the high RH region. It has been considered that the hysteresis phenomenon of cellulosic materials is resulted from the response delay caused by the collapse of nanoholes in the interfibrillar matrix as the internal water molecules exit as well as the delay of structural deformation during the adsorption process. To further explore the total hysteresis, the isotherm loops were calculated. Similar methods have been reported. The hysteresis loop of CNF was greater than that of CNF. The variations in the sorption hysteresis loops were attributed to structural swelling. Because the amorphous content of CNF is higher than that of the CNC, the extent of structural swelling in CNF is greater. These obvious differences between CNC and CNF are reasonable. Compared with nanocelluloses with cellulose I structure, nanocellulose with cellulose II structure have greater hysteresis loop. Cellulose materials with cellulose I structure first swell and then transform to cellulose II. Clearly, the observed differences in the hysteresis loops are due to the crystalline structure.

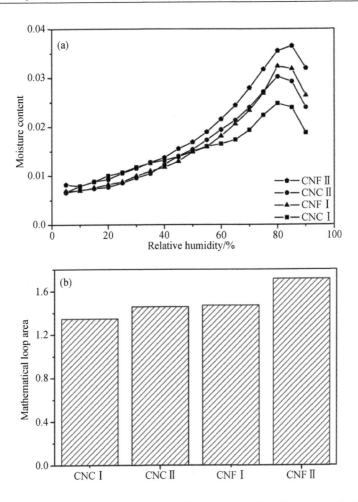

Figure 12.4 Sorption hysteresis (a) and mathematical loop areas (b) of four nanocellulose samples

To show the hysteresis, the incremental increase and the reduction of EMC vs. RH are shown in Figure 12.5. At the upper end of the RH range, the EMC increased during adsorption and decreased during desorption. However, in the other regions of the RH range, the EMC increase during adsorption and decrease during desorption was not obvious. Among the four nanocellulose samples, CNF II showed the most obvious increase and decrease in EMC [Figure 12.5(d)]. The EMC increase and decrease in the nanocellulose with the cellulose II structure were greater than those in the nanocellulose with the cellulose I structure. These differences are attributed to the differences in the crystalline structure.

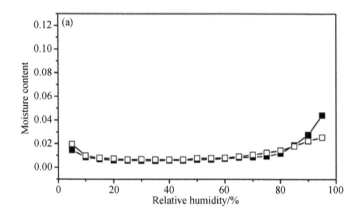

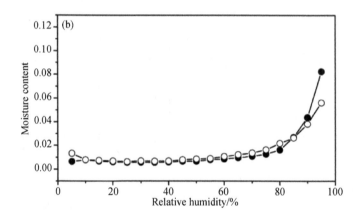

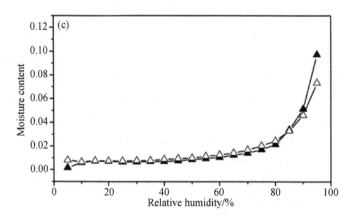

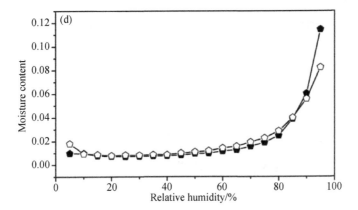

Figure 12.5 Incremental increase and the reduction of EMCs of CNC I (a), CNC II (b), CNF I (c), and CNF II (d) in the adsorption (solid points) and desorption (empty points) steps

12.3.3 Sorption kinetics

Figure 12.6 shows the average rate of sorption at various RHs during adsorption and desorption, in which the average rate at the specific RH was acquired by dividing the increment or decrement of EMC by time. The shape of the measured average rates for all four nanocellulose samples was an inverted V. This suggests that the average sorption rate is high at high RHs; however, in the low RH range during sorption and desorption, the average sorption rate is small. For RH higher than 60%, the average sorption rate of all four samples increased during absorption and desorption. In addition, the CNC curve for desorption is smoother than that for the adsorption. These differences in adsorption and desorption were further analyzed using the PEK model. The fitting results and corresponding t_1 and t_2 times are shown in Figure 12.7. Time t_1 and t_2 varied during adsorption and desorption process, especially at high and low RH; this variation in time t_2 was also found in microcrystalline cellulose. There are many differences between the characteristic time t_2 for adsorption and desorption. All these suggest that the adsorption and desorption are asymmetric processes. The cumulative moisture content related to the fast (MC_1) and slow (MC_2) steps in the water vapor sorption are shown in Figure 12.8. In adsorption, the CNC I and CNC II gained more mass during the fast step than the slow step [Figure 12.8(a) and 12.8(b)]. However, the CNF I and CNF II gained lesser mass during the fast step than the slow step at RH below 60% [Figure 12.8(c) and 12.8(d)]. This difference may be explained by the larger amorphous region with more sorption sites in the CNF. As mentioned above, the fast step is related to readily accessible sorption sites inside the surface, whereas the

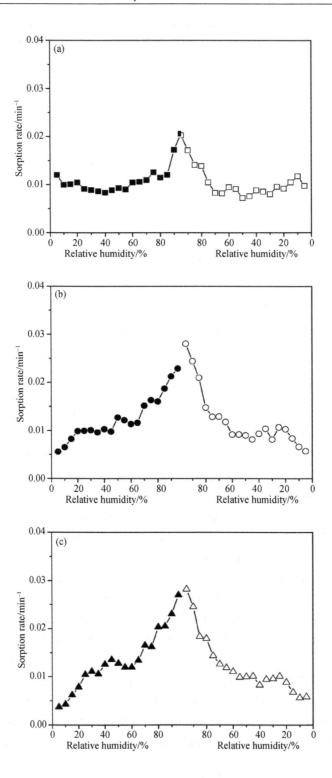

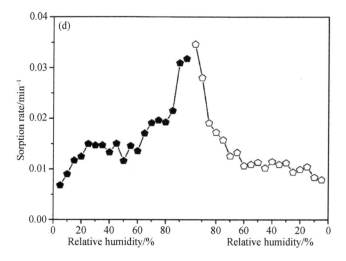

Figure 12.6 Average rates of sorption of CNC I (a), CNC II (b), CNF I (c), and CNF II (d) in the adsorption (solid points) and desorption (empty points) steps

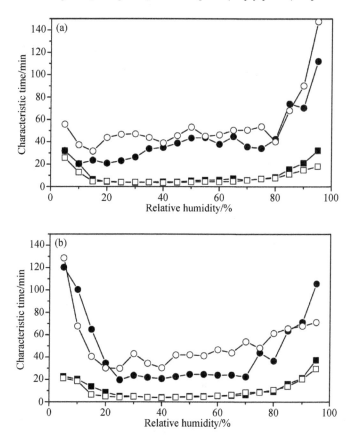

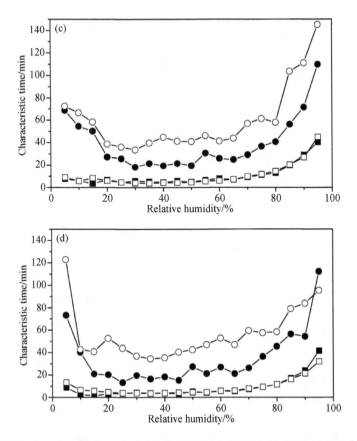

Figure 12.7 The characteristic times t_1 and t_2 of CNC I (a), CNC II (b), CNF I (c), and CNF II (d) in the adsorption (solid points) and desorption (empty points) steps

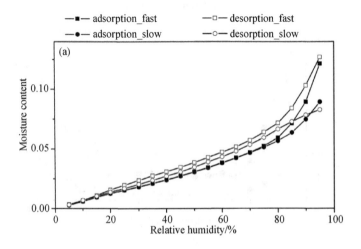

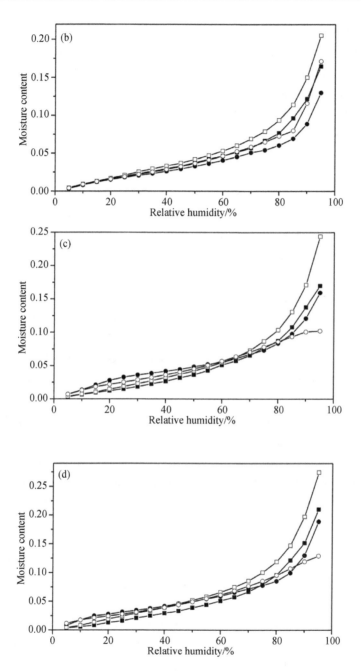

Figure 12.8 The cumulative moisture contents MC_1 and MC_2 of CNC I (a), CNC II (b), CNF I (c), and CNF II (d) in the adsorption (solid points) and desorption (empty points) steps

slow step is linked to the production of new water adsorption sites. Therefore, the CNF can gain more mass during the fast process than during the slow step. In desorption, there was also minor differences between the fast and slow steps. At 95% RH, the cumulative MC of both the fast and slow steps increased in the order of CNC I < CNC II <CNFI<CNFII. This is possibly because the fast and slow steps are related to the water absorption that increases in the same order.

12.3.4 The applicability of Kelvin-Voigt model

The Kelvin-Voigt model was introduced to estimate the modulus and viscosity of four nanocellulose samples; similar calculation methods have been previously reported in the literature. As shown previously, the kinetics of absorption or desorption consist of a fast and a slow step, with separate moduli and viscosities. In this study, the subscript in the modulus (E) and viscosity (η) was used to distinguish between the fast and slow step (subscript 1 for the fast and subscript 2 for the slow). The moduli and viscosities of the fast and slow steps for the four nanocellulose samples are shown in Figure 12.9-12.12. As expected, the variations in modulus and viscosity decrease with increasing RH, which is consistent with the effect of absorbed water on the plasticization of nanocellulose. There are obvious differences between the adsorption and desorption moduli E_1 and E_2, respectively. Similar variations are observed for viscosity. There is no previous report of the modulus of CNC and CNF calculated using Kelvin-Vogit model. However, our group has determined the modulus of CNC and CNF films using Instron 5582 testing machine and ASTM D638-10 standard (Qing et al. 2012) Here, we used the same procedures to determine the modulus values of CNC I, CNC II, CNF I, and CNF II which were (6.3±0.3) GPa, (5.3±0.2) GPa, (5.0±0.2) GPa, and (3.8±0.2) GPa. It is clearly that the modulus values of (23.9±1.8) GPa, (20.5±2.4) GPa, (19.2±1.8) GPa, and (18.0±4.4) GPa (at 50% RH) calculated using Kelvin-Vogit model in this study for CNC I, CNC II, CNF I, and CNF II are higher than those determined by tension method. Meanwhile, the modulus determined using tension method are acquired by applying external pressure on the nanocellulose, whereas the modulus calculated using Kelvin-Vogit model are based on an internal stress (the swelling pressure). Although the testing mechanism is different, the obtained modulus values are of the same order of magnitude. In addition, previous research has reported that the CNC film has modulus values ranging between 3 GPa and 25 GPa and another studies indicate the CNF film has modulus values ranging between 5 GPa and 21 GPa (Diddens et al. 2009; Lahiji et al. 2010). It is important to note that the modulus values calculated using Kelvin-Vogit model are in the range of modulus values quoted in the literature.

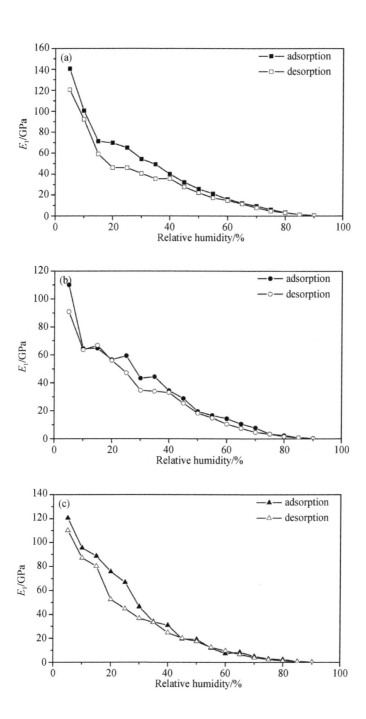

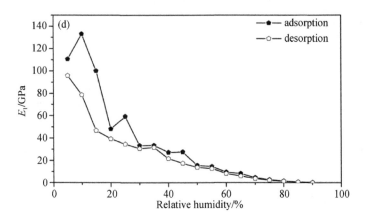

Figure 12.9 The moduli E_1 of CNC I (a), CNC II (b), CNF I (c), and CNF II (d) in the adsorption (solid points) and desorption (empty points) steps

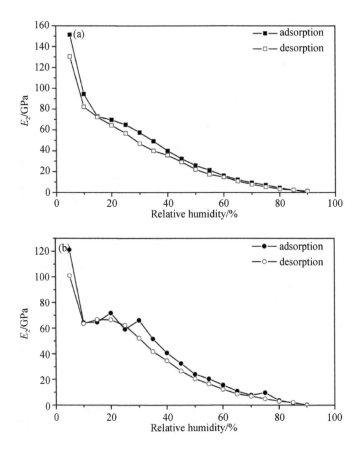

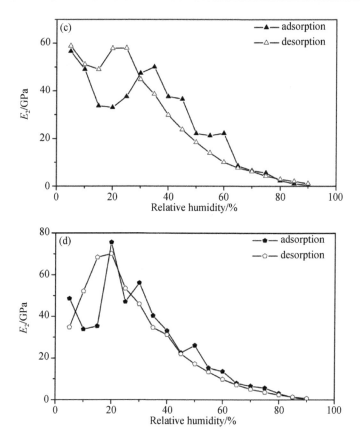

Figure 12.10 The moduli E_2 of CNC I (a), CNC II (b), CNF I (c), and CNF II (d) in the adsorption (solid points) and desorption (empty points) steps

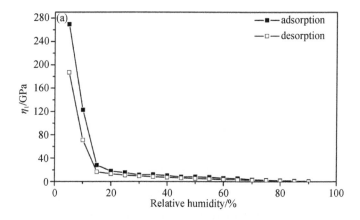

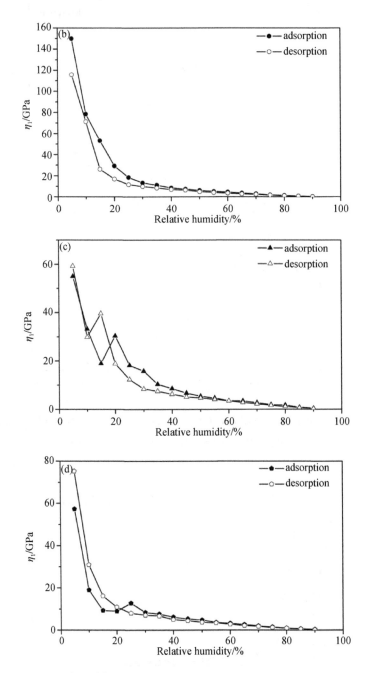

Figure 12.11 The viscosities η_1 of CNC I (a), CNC II (b), CNF I (c), and CNF II (d) in the adsorption (solid points) and desorption (empty points) steps

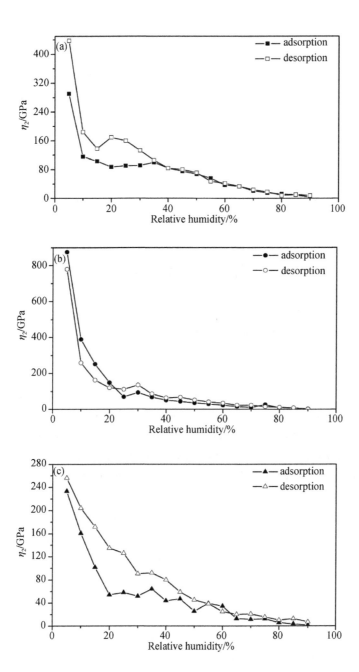

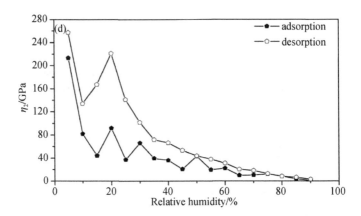

Figure 12.12 The viscosities η_2 of CNC I (a), CNC II (b), CNF I (c), and CNF II (d) in the adsorption (solid points) and desorption (empty points) steps

12.4 Conclusions

In this study, the differences in the water sorption behavior among four nanocellulose samples are discussed. Obvious differences were noted in the total running time, EMC, sorption isotherms, and the extent of sorption hysteresis. CNC I, CNC II, CNF I, and CNF II generated EMCs of 0.2142, 0.2856, 0.3321, and 0.3890, respectively, at RH of 95%. These differences were due to the amorphous content and crystalline structure of the samples. In addition, the dynamic sorption behavior of the four nanocellulose samples was analyzed by the PEK model. In the PEK model, the characteristic time and cumulative moisture content related to the fast and slow steps further showed the differences between adsorption and desorption. Subsequently, two Kelvin-Voigt elements were introduced to interpret the PEK behavior and calculate the moduli and viscosities of these four nanocellulose samples. The modulus values in this study are similar with the measured data in the literature; and they confirm that the DVS yields reliable data regarding the mechanical properties of nanocellulose.

References

AGRAWAL A M, MANEK R V, KOLLING W M, et al. 2004. Water distribution studies within microcrystalline cellulose and chitosan using differential scanning calorimetry and dynamic vapor sorption analysis [J]. Journal of Pharmaceutical sciences, 93 (7): 1766-1779.

ARGYROPOULOS D, ALEX R, MÜLLER J. 2011. Equilibrium moisture contents of a medicinal herb (Melissa officinalis) and a medicinal mushroom (Lentinula edodes) determined by dynamic vapour sorption [J]. Procedia Food Science, 1: 165-172.

AYTAC Z, SEN H S, DURGUN E, et al. 2015. Sulfisoxazole/cyclodextrin inclusion complex incorporated in electrospun hydroxypropyl cellulose nanofibers as drug delivery system [J]. Colloids and Surfaces B Biointerfaces, 128: 331-338.

AZEREDO H M C, ROSA M F, MATTOSO L H C. 2016. Nanocellulose in bio-based food packaging applications [J]. Industrial Crops and Products, 97: 664-671.

BALEA A, MERAYO N, FUENTE E, et al. 2017. Valorization of corn stalk by the production of cellulose nanofibers to improve recycled paper properties [J]. China Pulp and Paper Industry, 11 (2): 3416-3431.

BESSADOK A, LANGEVIN D, GOUANVÉ F. 2009. Study of water sorption on modified Agave fibres [J]. Carbohydrate Polymers, 76 (1): 74-85.

DENG Z, JUNG J, ZHAO Y. 2017. Development, characterization, and validation of chitosan adsorbed cellulose nanofiber (CNF) films as water resistant and antibacterial food contact packaging [J]. LWT - Food Science and Technology, 83: 132-140.

DIDDENS I, MURPHY B, KRISCH M, et al. 2009. Anisotropic elastic properties of cellulose measured using inelastic X-ray scattering [J]. Macromolecules, 41 (24): 9755-9759.

DRIEMEIER C, MENDES F M, OLIVEIRA M M. 2012. Dynamic vapor sorption and thermoporometry to probe water in celluloses [J]. Cellulose, 19 (4): 1051-1063.

DUMANLI A G. 2017. Nanocellulose and its composites for biomedical applications [J]. Current Medicinal Chemistry, 24 (5): 512-528.

ENDES C, CAMARERO-ESPINOSA S, MUELLER S, et al. 2016. A critical review of the current knowledge regarding the biological impact of nanocellulose [J]. Journal of Nanobiotechnology, 14 (1): 78.

FERRER A, PAL L, HUBBE M A. 2017. Nanocellulose in packaging: Advances in barrier layer technologies [J]. Industrial Crops and Products, 95: 574-582.

GAMELAS J A F, PEDROSA J, LOURENÇO A F, et al. 2015. On the morphology of cellulose nanofibrils obtained by TEMPO-mediated oxidation and mechanical treatment [J]. Micron, 72: 28-33.

GUO X, WU Y, XIE X. 2017. Water vapor sorption properties of cellulose nanocrystals and nanofibers using dynamic vapor sorption apparatus [J]. Scientific Reports, 7 (1): 14207.

HILL C A S, KEATING B, LAINE K, et al. 2012. The water vapour sorption properties of thermally modified and densified wood [J]. Journal of Materials Science, 47 (7): 3191-3197.

HILL C A S, NORTON A, NEWMAN G. 2010. Analysis of the water vapour sorption behaviour of Sitka spruce [Picea sitchensis (Bongard) Carr.] based on the parallel exponential kinetics model [J]. Holzforschung, 64 (4): 469-473.

HILL C A S, NORTON A, NEWMAN G. 2010. The water vapor sorption behavior of flax fibers analysis using the parallel exponential kinetics model and determination of the activation energies of sorption [J]. Journal of Applied Polymer Science, 116 (4): 2166-2173.

HILL C A S, RAMSAY J, LAINE K, et al. 2013. Water vapour sorption behaviour of thermally modified wood [J]. International Wood Products Journal, 4 (3): 191-196.

HILL C A S, XIE Y. 2011. The dynamic water vapour sorption properties of natural fibres and viscoelastic behaviour of the cell wall: Is there a link between sorption kinetics and hysteresis? [J]. Journal of Materials Science, 46 (11): 3738-3748.

HOENG F, DENNEULIN A, BRAS J. 2016. Use of nanocellulose in printed electronics: A review [J]. Nanoscale, 8 (27): 13131-13154.

ISA A, MINAMINO J, MIZUNO H, et al. 2013. Increased water resistance of bamboo flour/polyethylene composites [J]. Journal of Wood Chemistry and Technology, 33 (3): 208-216.

ISOGAI A. 2013. Wood nanocelluloses: Fundamentals and applications as new bio-based nanomaterials [J]. Journal of Wood Science, 59 (6): 449-459.

KACHRIMANIS K, NOISTERNIG M F, GRIESSER U J, et al. 2006. Dynamic moisture sorption and desorption of standard and silicified microcrystalline cellulose [J]. European Journal of Pharmaceutics and Biopharmaceutics, 64 (3): 307-315.

KACHRIMANIS K, NOISTERNIG M F, GRIESSER U J, et al. 2006. Dynamic moisture sorption and desorption of standard and silicified microcrystalline cellulose [J]. European Journal of Pharmaceutics and Biopharmaceutics, 64 (3): 307-315.

KEATING B A, HILL C A S, SUN D, et al. 2013. The water vapor sorption behavior of a galactomannan cellulose nanocomposite film analyzed using parallel exponential kinetics and the Kelvin–Voigt viscoelastic model [J]. Journal of Applied Polymer Science, 129 (4): 2352-2359.

KOGA H, NOGI M, KOMODA N, et al. 2014. Uniformly connected conductive networks on cellulose nanofiber paper for transparent paper electronics [J]. NPG Asia Materials, 6 (3): e93.

KOHLER R, ALEX R, BRIELMANN R, et al. 2010. A new kinetik model for water sorption isotherms of cellulosic materials [J]. Macromolecular Symposia, 244 (1): 89-96.

LAHIJI R R, XU X, REIFENBERGER R, et al. 2010. Atomic force microscopy characterization of cellulose nanocrystals [J]. Langmuir, 26 (6): 4480-4488.

LEE K, AITOMAKI Y, BERGLUND L, et al. 2014. On the use of nanocellulose as reinforcement in polymer matrix composites [J]. Composites Science and Technology, 105: 15-27.

LI F, MASCHERONI E, PIERGIOVANNI L. 2015. The potential of Nanocellulose in the packaging field: A review [J]. Packaging Technology and Science, 28 (6): 475-508.

LU Y, TEKINALP H L, EBERLE C C, et al. 2014. Nanocellulose in polymer composites and biomedical applications [J]. Tappi Journal, 13 (6): 47-54.

MADAMBA P S, DRISCOLL R H, BUCKLE K A. 1996. The thin layer drying characteristics of garlic slices [J]. Journal of Food Engineering, 29 (1): 75-97.

MALLADI R, NAGALAKSHMAIAH M, ROBERT M, et al. 2018. Importance of agriculture and industrial waste in the field of nano cellulose and its recent industrial developments: A review [J]. ACS Sustainable Chemistry and Engineering, 6 (3): 2807-2828.

MANDAL A, CHAKRABARTY D. 2011. Isolation of nanocellulose from waste sugarcane bagasse (SCB) and its characterization [J]. Carbohydrate Polymers, 86 (3): 1291-1299.

MANHAS N, BALASUBRAMANIAN K, PRAJITH P, et al. 2015. PCL/PVA nanoencapsulated reinforcing fillers of steam exploded/autoclaved cellulose nanofibrils for tissue engineering applications [J]. RSC Advances, 5 (31): 23999-24008.

MCCRYSTAL C B, FORD J L, RAJABISIAHBOOMI A R. 1997. A study on the interaction of water and cellulose ethers using differential scanning calorimetry [J]. Thermochimica Acta, 294 (294): 91-98.

MURR A, LACKNER R. 2018. Analysis on the influence of grain size and grain layer thickness on the sorption kinetics of grained wood at low relative humidity with the use of water vapour sorption experiments [J]. Wood Science and Technology, 52 (3): 753-776.

NECHYPORCHUK O, BELGACEM M N, BRAS J. 2016. Production of cellulose nanofibrils: A review of recent advances [J]. Industrial Crops and Products, 93: 2-25.

NEWMAN R H, DAVIDSON T C. 2004. Molecular conformations at the cellulose water interface [J]. Cellulose, 11 (1): 23-32.

OKUBAYASHI S, GRIESSER U J, BECHTOLD T. 2010. Moisture sorption/desorption behavior of various manmade cellulosic fibers [J]. Journal of Applied Polymer Science, 97 (4): 1621-1625.

OLSSON A M, SALMÉN L. 2004. The association of water to cellulose and hemicellulose in paper examined by FTIR spectroscopy. [J]. Carbohydrate Research, 339 (4): 813-818.

POPESCU C M, HILL C A S, CURLING S, et al. 2014. The water vapour sorption behaviour of acetylated birch wood: How acetylation affects the sorption isotherm and accessible hydroxyl content [J]. Journal of Materials Science, 49 (5): 2362-2371.

QING Y, SABO R, WU Y, et al. 2012. High-performance cellulose nanofibril composite films [J]. BioResources, 7 (3): 3064-3075.

RAMALHO J C, FORTUNATO A S, GOULAO L F, et al. 2013. Extraction and characterization of nanocellulose structures from raw cotton linter [J]. Carbohydrate Polymers, 91 (1): 229-235.

ROBERTKOHLER, CK R, BERNHARDAUSPERGER, et al. 2003. A numeric model for the kinetics of water vapor sorption on cellulosic reinforcement fibers [J]. Composite Interfaces, 10 (2-3): 255-276.

SHARRATT V, HILL C A S, ZAIHAN J, et al. 2011. The influence of photodegradation and weathering on the water vapour sorption kinetic behaviour of scots pine earlywood and latewood [J]. Polymer Degradation and Stability, 96 (7): 1210-1218.

SHI X, ZHENG Y, WANG G, et al. 2014. pH- and electro-responsive characteristics of bacterial cellulose nanofiber sodium alginate hybrid hydrogels for the dual controlled drug delivery [J]. RSC Advances, 4 (87): 47056-47065.

SIMON C, ESTEBAN L G, De PALACIOS P, et al. 2017. Sorption/desorption hysteresis revisited. Sorption properties of Pinus pinea L. Analysed by the parallel exponential kinetics and Kelvin-Voigt models [J]. Holzforschung, 71 (2): 171-177.

SUGIMOTO H, MIKI T, KANAYAMA K, et al. 2008. Dielectric relaxation of water adsorbed on cellulose [J]. Journal of Non-Crystalline Solids, 354 (27): 3220-3224.

SUN X, WU Q, LEE S, et al. 2016. Cellulose nanofibers as a modifier for rheology, curing and mechanical performance of oil well cement [J]. Scientific Reports, 6: 31654.

SUOPAJÄRVI T, LIIMATAINEN H, KARJALAINEN M, et al. 2015. Lead adsorption with sulfonated wheat pulp nanocelluloses [J]. Journal of Water Process Engineering, 5: 136-142.

TOPGAARD D, SÖDERMAN O. 2001. Diffusion of water absorbed in cellulose fibers studied with ^{1}H-NMR [J]. Langmuir, 17 (9): 2694-2702.

URAKI Y, MATSUMOTO C, HIRAI T, et al. 2010. Mechanical effect of acetic acid lignin adsorption on Honeycomb-Patterned cellulosic films [J]. Journal of Wood Chemistry and Technology, 30 (4): 348-359.

VELÁSQUEZ-COCK J, CASTRO C, GAÑÁN P, et al. 2016. Influence of the maturation time on the physico-chemical properties of nanocellulose and associated constituents isolated from pseudostems of banana plant c.v. Valery [J]. Industrial Crops and Products, 83: 551-560.

VOGT B D, SOLES C L, LEE H J, et al. 2005. Moisture absorption into ultrathin hydrophilic polymer films on different substrate surfaces [J]. Polymer, 46 (5): 1635-1642.

XIE Y, HILL C A S, JALALUDIN Z, et al. 2011. The dynamic water vapour sorption behaviour of natural fibres and kinetic analysis using the parallel exponential kinetics model [J]. Journal of Materials Science, 46 (2): 479-489.

XIE Y, HILL C A S, JALALUDIN Z, et al. 2011. The water vapour sorption behaviour of three celluloses: Analysis using parallel exponential kinetics and interpretation using the Kelvin-Voigt viscoelastic model [J]. Cellulose, 18 (3): 517-530.

XIE Y, HILL C A S, XIAO Z, et al. 2010. Water vapor sorption kinetics of wood modified with glutaraldehyde [J]. Journal of Applied Polymer Science, 117 (3): 1674-1682.

XIN G, LIU L, HU Y, et al. 2018. Water vapor sorption properties of TEMPO oxidized and sulfuric acid treated cellulose nanocrystal films [J]. Carbohydrate Polymers, 197: 524-530.

XU X, LIU F, JIANG L, et al. 2013. Cellulose nanocrystals *vs*. Cellulose nanofibrils: A comparative study on their microstructures and effects as polymer reinforcing agents [J]. Applied Materials and Interfaces, 5 (8): 2999-3009.

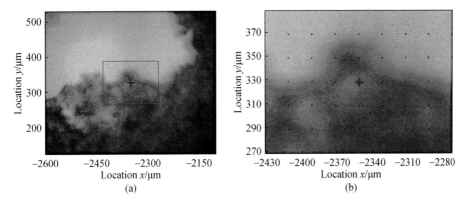

Figure 10.3 Visible light image of the cellulose nanofiber film. The red box indicates the position of the randomly selected mapping area including observation points (160 μm by 120 μm at 20 μm resolution, 63 observation points). In imaging acquisition mode, spectra are obtained sequentially from a series of observation points using point-by-point scanning

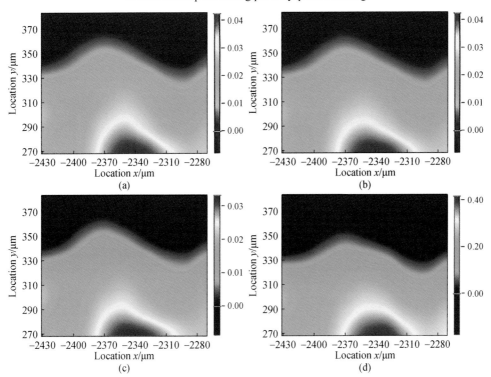

Figure 10.5 2-D micro-FTIR images (false color) of cellulose spatial distribution. Intensity scale appears on the right. Red locations indicate high concentration of cellulose; blue regions indicate low concentration

(a) This 2-D micro-FTIR image is generated using band of 2905 cm^{-1}; (b) This 2-D micro-FTIR image is generated using band of 1371 cm^{-1}; (c) This 2-D micro-FTIR image is generated using band of 1428 cm^{-1}; (d) This 2-D micro-FTIR image is generated using band of 1317 cm^{-1}

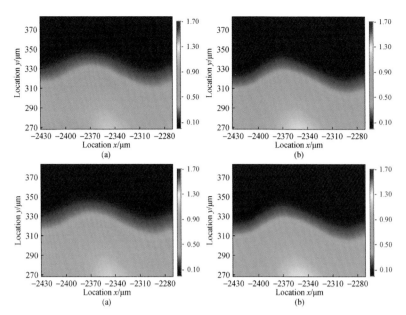

Figure 10.6 2-D micro-FTIR images (false color) of OH group distribution. Intensity scale appears on the right. Red locations indicate high concentration of OH group; blue regions indicate low concentration

(a) This 2-D micro-FTIR image is acquired at RH 0; (b) This 2-D micro-FTIR image is acquired at RH 30%; (c) This 2-D micro-FTIR image is acquired at RH 60%; (d) This 2-D micro-FTIR image is acquired at RH 90%

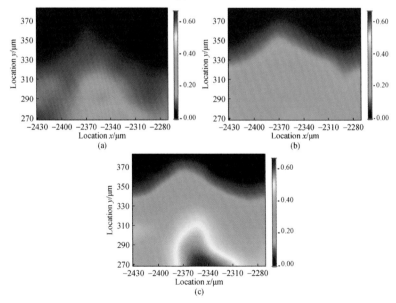

Figure 10.7 Difference 2-D micro-FTIR images (false color) characterizing adsorbed water distribution. Intensity scale appears on the right. Red locations indicate high concentration of adsorbed water; blue regions indicate low concentration

(a) This difference 2-D micro-FTIR image characterizes the water adsorbed below RH 30%; (b) This difference 2-D micro-FTIR image characterizes the water adsorbed below RH 60%; (c) This difference 2-D micro-FTIR image characterizes the water adsorbed below RH 90%